T0341214

# Sparks from the Spirit

# Sparks from the Spirit

## From Science to Innovation, Development, and Sustainability

Yongyuth Yuthavong

Pan Stanford Publishing

*Published by*

Pan Stanford Publishing Pte. Ltd.
Penthouse Level, Suntec Tower 3
8 Temasek Boulevard
Singapore 038988

Email: editorial@panstanford.com
Web: www.panstanford.com

**British Library Cataloguing-in-Publication Data**
A catalogue record for this book is available from the British Library.

**Sparks from the Spirit: From Science to Innovation, Development, and Sustainability**


ISBN 978-981-4774-57-4 (Hardcover)
ISBN 978-1-315-14599-0 (eBook)

*To Onchuma, Namon, and Rasa*

# Book Reviews

*This book has an important message for anyone concerned about the future of mankind and our planet. Elegantly and concisely, Prof. Yuthavong traces our efforts through millennia to understand this world and ourselves and science's pivotal role therein. He notes science's evolving nature (we still don't have a single theory or a theory of everything) but rightly emphasizes that scientific advances and innovations in and of themselves will not be enough for a sustainable future; they must be coupled with a better understanding of human nature and human society (where science could also be of great help) and supported by good policies and good governance. He also warns us against some of the current practices that could have disastrous consequences and highlights the importance of sustainability in all our current and future development endeavors.*

**Dr. Mohamed ElBaradei**

Nobel Peace Prize winner, Former Vice President of Egypt, and Director General Emeritus, International Atomic Energy Agency, Austria

*Prof. Yuthavong guides us through an absorbing panorama of human progress ignited by sparks from the spirit of science throughout all ages and across all continents. He writes with humility and deep conviction that scientific and technology innovations tempered by social justice and common humanity will lead humankind to sustainability.*

**Academician Dato' Ir. (Dr.) Lee Yee Cheong,**

Chairman, UNESCO International Scientific, Technology and Innovation Centre for South-South Cooperation

*Prof. Yuthavong revitalizes the concepts of science and innovation in* Sparks from the Spirit*. He displays a passionate love affair with science and lights a match that will surely enable any nonscientist to find his or her way in a dark room. People like me will be enticed to reach for more books about science.*

**Mechai Viravaidya,**

Population and Community Development Association, Thailand

*This most informative and enthralling narrative, anchored with a sound knowledge of science and the highest levels of public policy decision making, authoritatively links creativity, wonder, exploration, inventiveness, and development with the central role of science, sparking the quest for sustainability since the dawn of civilization. The book is well structured. The chapters can be read sequentially or as stand-alone chapters with specific themes and thrust and with historical and contemporary examples.* Sparks from the Spirit *is an invaluable resource for arousing interest and igniting action, particularly in decision makers in governments, business, and academia, toward a sustainable future and beyond. The book provides many inputs into various aspects of sustainability for both society and the environment.*

**Nay Htun**

Founder and Hon. Patron, Green Economy Green Growth Association, Myanmar

*For most laypersons, science is at best worthy but dry and boring or at worst the root of our major problems. Where* Sparks from the Spirit *excels is in pointing out the beauty of science and its role in inspiring a sense of wonder and the urge to explore. It highlights the open-ended nature of the enterprise, always growing, always embracing new dimensions. The book shows that science is not just a European invention but has major contributions from Asia (especially China and India) and Arab/Islamic civilisations. Through numerous examples, the book also conveys a rich array of scientific discoveries benefitting humanity.*

**Sir Gustav Nossal,**

Former President of the Australian Academy of Science,
former President of the International Union of
Immunological Societies, and
Australian of the Year 2000

# Contents

# Preface

This book traces the links between science, which is the human quest for knowledge, and sustainable development. Science contains all knowledge about nature, which we obtain by asking questions, observing, and experimenting in attempts to find answers and make rational conclusions. It is a common heritage of humankind, stemming from curiosity, imagination, and experience in seeking explanations to understand all things around us. The spirit of scientific enquiries leads further, from an understanding of our problems to solutions that enable us to survive and develop. The "sparks" originating from the spirit of science lead to innovation and are essential for achieving sustainable development, which benefits us all, including the poor and the disadvantaged, whose needs are most acute. The book points out that common people as well as government and business have big, potential roles to play in sustainable development, a major spark with one of its roots in the spirit of science.

The book is intended for a general, especially young readership. I hope it appeals to those curious about the nature of science and the benefits it brings, together with possible pitfalls it may have. It begins with the role that science plays in contributing to the quality and content of learning of individuals. It then explores the effects of science on societies, highlighting the opportunities it gives to newly developing societies that can overcome obstacles in order to achieve sustainable development. It is written in nontechnical language, with illustrations, case studies, practical examples, and historical perspectives. It is intended to create both an understanding of science and an appreciation of the role it plays in human development. I am a professional scientist with a broad interest in social and economic development, especially of developing countries. I have combined my technical background as a research scientist and administrator with my expertise in policy and practice derived from experience as former Minister of Science and Technology and former Deputy Prime Minister of Thailand.

**Yongyuth Yuthavong**

2018

# Background and Acknowledgments

*Sparks from the Spirit* takes up the theme of the nature and purpose of science, with the aim of communicating with general readers, including young people who may be curious about the work of scientists and their impact on society. It originated from the talk entitled "What Is Science For?," given as the 2011 annual invitation lecture in honor of Prof. Dr. Puey Ungphakorn, former president of Thammasat University. He was recognized by UNESCO recently, on the occasion of his 100th anniversary, as one of the world's most important people for his impeccable ethics and his role in national development. Important for me personally, he was my beloved uncle, who deeply influenced my life and career. The talk was subsequently published as a book in Thai under the same name, with support from the Crown Property Bureau of Thailand. My participation in the work of the Thailand Sustainable Development Foundation has provided me with further ideas on the sparks from the spirit of science, especially those concerning sustainable development.

I would like to thank Dr. Chirayu Isarangkun Na Ayuthaya for his continuing support and encouragement. Thanks are due to Thamrong Prinyakanit (GPEN) for cartoon drawings, which greatly enliven this book. Phil Shaw has been very helpful in editing this book. Thanks are also due to Dato Lee Yee Cheong for advice and help in various stages of the book, and to Dr. Mahamed ElBaradei, Dato Lee Yee Cheong, Mechai Viravaidya, Dr. Nay Htun, and Sir Gustav Nossal for their reviews of the book. I would also like to thank Archana Ziradkar and Jenny Rompas, who have given much help in the publishing process.

# Synopsis

A long-lasting activity—not just over a few years or a few generations but from the dawn of civilization—must have a spirit. The spirit encompasses the essence of the activity, which engages all concerned in a continuing process handed down from generation to generation, with increased and more refined contents. The spirit of science never dies. When we talk of the spirit of a nation, we believe that the nation will live on and flourish, even though it may encounter difficulties from threats of war or domination by others. When we talk of the spirit of a person, we believe that his or her character lives on, even though he or she may have departed from the world. The spirit of science is not only a continuing entity but also one that has grown and evolved over past millennia.

As a body of knowledge about nature obtained through experiments and observation, science began as early as, or perhaps even earlier than, the moment when humans discovered fire and learned to make various tools. Wonders about nature, curiosity, and the sense of exploration must have arisen with the dawn of human evolution. All ancient civilizations had "science" as knowledge to guide their everyday lives, lumped together at first with other branches of knowledge. Early science evolved in a few places, including Greece, Egypt, and the Arab world, into distinct areas of astronomy, physics, medicine, and alchemy, which later gave rise to chemistry. Civilizations of the East, notably China and India, gave rise to a parallel development of science, some of which later merged with that of Europe and the Middle East through trade and other contacts. The age of Renaissance from the 14th century saw the flowering of science in Europe. Application of scientific discoveries—the sparks from the spirit of science—in production and services gave rise to a major transition in Europe, called the Industrial Revolution. The spirit of science and its sparks spread worldwide with growing momentum, helped by international trade, industry, and other human interactions, to the present age of the post–Industrial Revolution, with major advances in information technology, technology of materials and their production, and technology of health and life, down to the details of human cognition. Such advances have been mostly exploited successfully in the developed countries, which already have the infrastructure and people who are ready to move forward. However, many developing countries, where the majority of people are still struggling just for day-to-day survival, have not had significant benefits from scientific advances in the past. They now enjoy better opportunities given by science-based technologies, including greater connectivity, to help along the process of development. International effort is needed, in addition to indigenous efforts, to make sure that developing countries can fully reap the benefits of science and technology.

The spirit of science that gives rise to its success is due to contributions of scientists, who must be nurtured over several years before they can mature to be productive, together with many partners. Curiosity, and the creativity that comes with it, is the most important character for productive science. Curiosity is innate in every child, but it has a tendency to wane because of various factors as we grow older, including inadequate societal encouragement, distractions, and even discouragement as a consequence of peer pressure to conform to the norms in many societies. The urge to invent is also innate in many children, and again it tends to fade away once they grow up, for lack of sufficient encouragement, opportunities, and distractions. It is therefore important to have an education system that encourages thinking, learning, and doing things by oneself. Such an education system integrates basic science with applied science and technology, together with other aspects of learning, hence both nurturing the spirit of science and enhancing the sparks from this spirit. While such a system may already be largely in place in many developed countries, it is lacking in most developing countries.

The sparks from science include innovations and various applications, contributing to economic benefits and human wellness, in turn leading to development and sustainability. The benefits to many aspects of human life are so amazing as to invite comparison with fairy tales. However, access to these benefits is not universal, with those at "the bottom of the pyramid" often missing out. The imbalance of benefit sharing threatens the stability of society. Consideration should therefore be made, in public policy and general public consciousness, to ensure that the benefits of science and technology reach all members of society. Likewise, the poorest nations should no longer be neglected and denied the benefits of science, now that the world is more interconnected than at any time in history.

In spite of benefits from science and technology, we must be cautious of potentially adverse effects, many of which require ethical, social, and legal considerations. Products of science and technology, such as powerful weapons, have been used to destroy lives, although some are justified politically as deterrents. Information technology, which has revolutionized almost all aspects of our lives, has been used as a powerful tool of sabotage. Looming around the corner, the use of stem cells and genetic modification of humans and animals, with tantalizing promises for health and production of food, stand the risk of becoming misappropriated by ill-intentioned people or simply for change of life forms without due consideration of long-term implications.

The myriad of benefits that science brings come mainly from technology associated with it and not directly from science itself. Science provides the tools of technology, essential for production and services, and yet these benefits do not constitute the direct purpose of science, which is to obtain knowledge concerning nature. This can be extended further to include knowledge about our own inventions. Technology provides the know-how and leads to products and services, which we invent, not just by trial and error, but also by reasoned maneuver based on scientific knowledge. The new millennium is seeing most nations engaged in efforts to achieve goals of sustainable development, including the end of poverty, universal health coverage, and equal opportunities for all regardless of gender, age, or origin of birth. Such goals should be achievable with the aid of science and technology. Contributions to sustainable development would be the big sparks of the spirit of science for the new millennium and thereafter.

# Guide to the Contents of the Book

This book is written for those who are broadly interested in the nature of science and its sparks as benefits for society, using mainly nontechnical language, and it is hoped that this book will help bridge the gap between science and the society at large. For this book, the "spirit" is the spirit of science and the "sparks" are benefits or, in some cases risks and dangers, derived from science. Chapter 1 ("The Spirit of Science") introduces the reader to the essence of science, which has a spirit born in all of us, growing in strength in some and decaying in others. Chapter 2 ("The Spirit Has Sparks") points out that for those whose spirit has grown strong, sparks can be produced in the forms of discoveries, innovations, and development. Chapter 3 ("Nurturing the Spirit, Enhancing the Sparks") makes a case for growing the spirit of science and enhancing the sparks from it through collaboration with people who can help in bringing science benefits to society. Chapter 4 ("Cultivating the Brain") gives importance to the role of education in growing the spirit of science and enhancing the sparks from it. Chapter 5 ("Creating Sparks from the Spirit") points out that creativity and inventiveness are needed to produce sparks from the spirit, not just knowledge and skills. Chapter 6 ("What Is Science For?") examines the dual purpose of science in acquiring knowledge and obtaining the benefits derived from it. Chapter 7 ("Present and New Challenges") asks what challenges remain and what the new challenges for science and its sparks are. Chapter 8 ("Addressing the Base of the Pyramid") emphasizes the use of science for the benefits of the disadvantaged and deprived populations of the world. Chapter 9 ("Dangers and Risks in the Sparks") raises ethical, social, and legal issues in science and technology, which can produce both unintended risks and dangers and have been used in destructive manners. Chapter 10 ("Sparks for Sustainable Development") explores the role of science in contributing toward sustainable development, and Chapter 11 ("Moving Beyond Sustainability") looks further to its role in the future, where issues beyond mere sustainability loom large for humanity.

This book can be read from beginning to end in a linear manner. However, selected chapters can also be read independently by readers who want to visit particular aspects. For example, those who would like to see the links between science and education can start at Chapter 4, and those who are interested in the potential role of science for relief of poverty and people at the base of the pyramid can go to Chapter 8. Those interested mainly in the ethical, social, and legal aspects of science and technology can go directly to Chapter 9. It is hoped that this book will help in an understanding of the spirit of science and its sparks and in our common search for a sustainable world and hopefully the bright future beyond.

## Chapter 1

# The Spirit of Science

*I keep six honest serving-men*
*(They taught me all I knew);*
*Their names are What and Why and When*
*And How and Where and Who.*

—Rudyard Kipling, *The Elephant's Child*, 1902

Science is the body of knowledge about nature that we obtain from asking questions concerning things around us and finding answers through observation and experiments. The main driving forces behind science are curiosity, the sense of wonder, and the quest for knowledge through exploration. The spirit indeed exists in all children, only to be obliterated in many as they grow up without due encouragement by society. The spirit of science helps build the ability of individuals to face and solve their own problems in daily life and help society in making sound decisions and exploiting the fruits of science. Guiding humans away from awe and superstition dominating ancient societies, science has grown independently in many civilizations. Modern science now encompasses vast knowledge about nature, from the smallest components of matter to stars and galaxies, with the tendency to integrate various fields of specialty. Spheres of human experience about the arts, literature, and history also contribute to, and have significant contribution from, science. By nature, scientific knowledge, covering both

*Sparks from the Spirit: From Science to Innovation, Development, and Sustainability*
Yongyuth Yuthavong

ISBN 978-981-4774-57-4 (Hardcover), 978-1-315-14599-0 (eBook)
www.panstanford.com

subject matter and understanding, is always open to more new knowledge and is therefore always growing. Moreover, its success has encouraged other subject areas, such as human, social, and political studies, to emulate its approach. Human and social sciences are indeed crucial for the success of applying the fruits of science and technology for sustainable development.

## 1.1 The Fires within Us

Inside us, we are all still children. We are curious about things around us, just as children are. We are generally curious about what is happening, how and why it happens, and what will follow. Our curiosity covers various things in day-to-day life, our livelihood, our neighborhood, and our world. On most occasions, the routines of daily life do not allow us to follow our curiosities, except where they concern us personally. Children, however, follow their curiosities more freely, although, of course, they are less skilled in finding answers about them than adults. They ask seemingly simple and straightforward questions like why, how, and what for. Where did the first human come from? If a parent answers "A human-like ape perhaps," then more questions certainly arise on where the human-like ape comes from and how it can give rise to a human anyway. Other questions like why people die, why ice melts in our hands, why the sun shines, and why the moon keeps following us wherever we walk are typical of what engages young minds. Come to think of it, these are also the type of questions we ask ourselves, even as grown-ups. We certainly would like to know the answers; only we are so busy with other things in our everyday lives that we never have the time to find out. Some of us got to know the answers from our parents or from the classroom, and some from the Internet, but some questions remain unanswered. They are left as puzzles in our mind—like debts we never get to pay. The inborn sense of curiosity stays with us all our lives, no matter whether we have the time or the opportunity to satisfy it.

So what good is curiosity? Why do we need to retain it past our childhood? Worse still, would curiosity get us into trouble, like children who do not listen to parents about the dangers of wading into deep water, playing with unknown animals, or going too near cliffs? Indeed there are many tales warning us of hazards of following our curiosity (Box 1.1). Curiosity can therefore be considered both a human weakness and a human virtue. It is a weakness because it exposes us to the risk of the unknown, with potentially bad outcomes, like drowning on plunging into a deep well, being bitten by an unfamiliar dog, or falling from a cliff on venturing too far. Grown-ups know these risks well, and warn us against them and against following the curiosity that is a weakness in these examples. "Curiosity killed the cat" is a warning we often hear. However, despite the risks involved, curiosity can also be considered a virtue as it encourages us to search for new experiences, to explore new possibilities, and to imagine the consequences of following new actions. Learning is enriched as we become engaged in creative ideas and actions triggered by our curiosity. Curiosity-driven learning is an active process that helps us understand and construct our view of the world around us. Indeed, progress in science would not be possible without curiosity-driven investigation. Therefore, while we should admit that there is a risk involved in following curiosity without due caution, it is definitely a virtue worth following. "Curiosity killed the cat, but satisfaction brought it back"—the satisfaction of fulfilling our basic sense successfully, as the rejoinder goes.

**Box 1.1** Tales of Curiosity

Is curiosity a human weakness or a virtue? Or both? Whatever it is, it is certainly a common human character. Many tales from around the world tell of how the call of curiosity leads to often seemingly disastrous consequences but finally to "living happily ever after." Take "Sleeping Beauty," a fairy tale by the Grimm Brothers. Curious about the pointed spindle of the spinning wheel, Princess Aurora falls victim to the bad fairy Maleficent by touching the spindle, pricking her finger, and falling asleep for a hundred years. Numerous stories concern how someone, being forbidden to do something, succumbs to the temptation of doing just that, for being curious as to what would happen. Here are stories from two different parts of the world about curiosity on what is in the box, the opening of which is forbidden (Fig. 1.1).

In a Thai folk tale, Jantakorop was a prince who studied with a *rishi* (a holy man) in the forest. To reward Jantakorop for his success in the study, the rishi gave him a small box as a present but forbade him to open it before he reached his palace. On the way to the palace, Jantakorop succumbed to the curiosity of wanting to know what was in the box, and opened it. A beautiful lady, named Mora, came out, and they fell in love. Traveling together further, they encountered a bandit, who wanted Mora for his wife. Jantakorop fought with the bandit, and called for a knife from Mora. Attracted by the bandit, however, Mora gave the knife to the bandit and he killed Jantakorop.

According to Greek mythology, Pandora was the first woman on earth, created by the gods, each of whom gave her a gift (*pandora* means *one who bears all gifts*). Beautiful and seductive, she was actually created as a punishment for humans by the great god Zeus, who was angry that Prometheus (the creator of mankind) stole fire against his will and gave it to the humans. In a cunning scheme, Zeus sent Pandora as a gift to be married to Epimetheus, brother of Prometheus. Although warned by Prometheus not to accept any gift from the gods, Epimetheus was completely charmed by Pandora and agreed to marry her. Now, Zeus gave Pandora a box with a big key as a wedding gift but instructed her never to open it. Pandora promised as told but was tormented by curiosity as to what was in the box and why she should not open it, since it was a gift for her and she would like to know its contents. Curiosity finally got the better of her, and when Epimetheus was out of sight, she took the key and started to open the box but changed her mind three times. Eventually she succumbed to curiosity and opened it. Instead of jewelry, gold, or other treasures, Pandora was unpleasantly surprised that all that came out were evil creatures like disease, poverty, sadness, and death in the form of tiny moths, which stung her badly. She hurriedly shut the box, ignoring a voice from inside asking to be let out also. Epimetheus heard Pandora's cry of pain and came in to see what happened. They opened the box once more and found that all that remained was Hope. Thankfully, Hope fluttered out of the box like a dragonfly, touching and healing all the wounds created by the evil creatures.

These tales of curiosity seem to send a message that it has bad consequences. Sleeping Beauty was asleep for a hundred years, Jantakorop was killed by the bandit, and Pandora was stung by evil creatures. Many people use these tales to warn children, and even grown-ups, not to be too curious, and there are indeed good reasons, since there are dangers and risks of the unknown. However, a better message for children, who are mostly naturally curious anyway, should be that they should follow their sense of curiosity with caution.

By the way, the ends of these stories of curiosity are not too bad. Princess Aurora did get to meet her prince, albeit about a century younger than her. Jantakorop was revived by God Indra and got to marry a good lady, while Mora was condemned to roam the forest forever in the form of a gibbon, looking for a husband. As for Pandora, she not only was redeemed by Hope but is also now famous by the term "Pandora's Box."

**Figure 1.1** Learning from tales of curiosity: "Pandora," "Jantakorop," and "Sleeping Beauty."

Together with a sense of curiosity, we also have a sense of wonder. Recall the wonder we felt when we saw a good magic show, a huge waterfall, or simply a star-studded sky on a clear night. We appreciate the spectacle of something mysterious, something bigger than us, and we want to know, see, and understand the object of our wonder more. Sometimes the object of our wonder fills us with awe, astonishment, and fear. These feelings lead many people toward a religious experience. However, the sense of wonder, so strong when we were young, tends to diminish as we grow older, to our own detriment. In the words of Rachel Carson [1],

> A child's world is fresh and new and beautiful, full or wonder and excitement. It is our misfortune that for most of us that clear-eyed vision, that true instinct for what is beautiful and awe-inspiring, is dimmed and even lost before we reach adulthood.

It might be thought that we experience the sense of wonder only when we meet unusual things, such as rare works of art, exquisitely beautiful sceneries, or meteor showers. Certainly, such unusual encounters can invoke the sense of wonder, but they are not the exclusive origins of this sense. In fact, if we are prepared to extend our perception of the world and look for new ways of seeing, hearing, and feeling, we can experience the sense of wonder in everyday life, with common surroundings and things around us. A magnifying glass, for example, can reveal so many unfamiliar and unexpected life forms under the grass or in a pond. Sea shells are of so many different sizes, colors, and forms as to invoke wonder on how they were formed, what the creatures that used them as protection look like, and why other creatures, such as fish or indeed we humans, do not have such shells. A thunderstorm not only makes us feel thankful to be safe in our homes but also invokes a sense of wonder at the fierce and gigantic force of nature. We also wonder at some facts revealed to us, such as the fact that DNA (short for deoxyribonucleic acid), the genetic material in our cells, would if stretched out into a single fiber extend as long as 60 times the distance to the sun! Like curiosity, the sense of wonder is surely something we should treasure, something that enriches our lives and makes us thankful for the experience and realize the vastness and complexity of the world, and indeed the universe around us.

The senses of curiosity and wonder spur us on to undertake another activity, again strong when we were children—exploration. However, the urge to explore does not always come from curiosity and wonder but can come from such motivation as potential benefit or simply just to achieve the goal of exploration. "Because it's there" is the famous answer of mountaineers in response to the question of why they want to get to the summit. The urge to explore remains with us all our lives, although again in some people it tends to be less strong as they grow older, for reasons similar to those already discussed for the senses of curiosity and wonder. However, we never lose these three senses completely, and they seem to be the main characteristics that distinguish us from animals. True, apes and other animals also have senses of curiosity and exploration to different degrees, but they are not as strong and, as we will discuss, not followed in an intense, continuous, and systematic manner as in humans. As for the sense of wonder, chimpanzees are reported to have it, but this is probably rare in other species and only rudimentary even in chimpanzees. Humans are the only species that has these senses to great depths and in extensive ways.

Many explorers in history followed their urge and went on to make great discoveries. Consider the history of naval explorations. Explorers like Christopher Columbus, Vasco da Gama, Ferdinand Magellan, and James Cook made important discoveries of lands previously unknown to the Europeans, paving the way for expansion of colonial powers to areas that until then were inhabited only by indigenous populations. Christopher Columbus (1450–1506) was an Italian explorer searching for a new route to Asia and the Indies under the auspices of the Catholic monarchs of Spain and instead found the New World. Vasco da Gama (1460–1524), a Portuguese explorer, was the first European to reach India by sea. Ferdinand Magellan (1480–1521), a Portuguese explorer, almost succeeded in the first circumnavigation of the globe, except for his death in the Philippines. James Cook (1728–1779) was a British explorer who first made European contact with Australia and the Hawaiian Islands and made the first circumnavigation of New Zealand. Great naval explorations, however, were not only made by the Europeans. Well before the arrival of Columbus in the Americas and that of da Gama in India, a naval commander from China named Zheng He (1371–1433) made several voyages from China to various places in Southeast and East Asia, the Middle East, and Africa with huge armadas. The first of seven voyages, for example, consisted of a fleet of 317 ships holding 28,000 crewmen. Although the routes of these travels had been known earlier, expeditions of such large magnitude were not, and technologies for making and navigating such gigantic ships were legacies of his achievements. For example, the sizes of the ships were unmatched by any wooden ships ever recorded in history, with lengths of up to 120 m and widths of up to 50 m. By comparison, the length of Columbus's flagship Santa Maria was only one-fourth of the lengths of Zheng He's ships. These accounts show that historical people from both the East and the West undertook great explorations so as to discover new places. The main motivations for these explorations were potential material benefits, in terms of trade, treasure, and new opportunities, plus in some cases the spread of religious beliefs. However, the curiosity to know what lies beyond their familiar places and the wonder of strange lands and seas must surely have been a part of their motivation. More importantly, their discoveries of faraway lands, with exotic plants, animals, valuable merchandise, and unfamiliar people, certainly aroused the sense of curiosity (and greed) in the people who got to know the stories, triggering the eventual colonization of these new lands.

Exploration usually consists of two main activities, survey and investigation. Survey is a broad undertaking, where the overall features of the place or object of exploration are assessed and the main characteristics noted. Investigation is a more detailed study of the place or object, where both the overall aspects and the specific features of interest are probed so as to obtain information about say the people, animals, and plants inhabiting the land. Survey and investigation eventually yield knowledge about the place or object and, more importantly in many cases, an understanding of its various features in detail. Understanding is often something that comes much later than the information obtained. It is this information and, more so, understanding that satisfy our senses of curiosity and wonder, and in many cases, when we cannot yet understand our object of exploration, our sense of wonder is further reinforced. We all experience these senses when we visit a new place or when we see something new and interesting.

## 1.2 Spirit Arising from the Fires

The fires of curiosity, sense of wonder, and urge to explore never die down for most of us, even though they may become dimmer in some. They combine to form a lasting spirit—the spirit of science—encompassing the activity that engages all concerned over the millennia of human civilization. The body of knowledge called science, with the spirit arising from these fires, was at first not clearly distinguished from other spheres of knowledge, but it gradually gained a character of its own in a continuing process handed down from generation to generation, with increased and more refined contents. Early human knowledge, first passed on through oral tradition, was preserved and expanded through the development of writing. This early knowledge was a mix of observation of nature, social codes, religion, and mysticism. For example, astronomy, the study of objects and phenomena concerning stars and celestial bodies so as to understand and predict their nature and motion, was at first not distinguished from astrology, which uses the positions and motions of the stars as the basis for prediction of future events, until around the 18th century, when they were gradually regarded as different disciplines. Alchemy, the art of dealing with substances with the aim of turning base metals into noble ones like gold and silver and finding the elixir of life conferring youth and longevity, was a mix of spiritual practice and diligent observation and experimentation that partly gave rise to modern chemistry and medicine.

Successive early civilizations in Mesopotamia, situated largely between the Euphrates and Tigris Rivers (now Iraq and parts of Iran, Turkey, and Syria), made advances in agriculture, animal domestication, tools and weaponry, sailing, and irrigation. The region of Sumer in south Mesopotamia, in the fourth millennium BCE, saw the first city settlement, the invention of the wheel, and the invention of writing (which also developed in other places, including Egypt, the Indus Valley, China, and Mesoamerica). It is in this area that people turned from hunter-gatherers into settler-dwellers of land and cities. Trade and communication enabled development of different skills and occupations, aided by learning. In other parts of the world parallel development occurred soon thereafter. The Chinese civilization, which goes back some 5000 years, saw the birth of important inventions like paper, printing, the compass, and gunpowder. The Indus civilization, situated in the northwest Indian subcontinent, also started around the same period and saw the development of brick housing, the water supply and sanitation system, metallurgy, and urban planning. The ancient Egyptian civilization arose in a series of kingdoms, noted by advances in quarrying, surveying, and construction techniques that gave rise to temples and pyramids, agriculture, medicine, and mathematics.

These early civilizations heralded the beginning and development of the arts, mathematics, astronomy, medicine, and various inventions. Although they were derived partly from the demand for making a living and staying together as a society, a main driving force for these achievements was undoubtedly the spirit arising from the combination of curiosity, wonder, and exploration. However, this early spirit was influenced by religious beliefs and teachings derived from faith and myths regardless of rationality. The spirit of science emerged from this early broad spirit and took a clearer form separated from mysticism and superstition, starting with the rise of the ancient Greek civilization. Influenced by earlier civilizations, the ancient Greeks, and to a lesser extent the ancient Romans, were interested in understanding the world and everything around and about them, including the stars and how life originated. These interests went deeper than mastering the skills in trade, craftsmanship, metal working, construction, and warfare, but they went all the way to how and why the world started and life began. Although the ancient Greeks also had their myths and many natural events were ascribed to the gods, they had the unique spirit of enquiry that would gradually form the spirit of science handed down from generation to generation to the present day.

The spirit of science started first with philosophical enquiries into the nature of reality, existence, change and motion, and nonliving and living states. This gradually turned to exploration and observation of nature in order to answer these questions. Examples of the early pioneers of the spirit include Pythagoras (570–495 BCE, mathematician and philosopher), Plato (428–348 BCE, natural philosopher, or philosopher who studied nature and the universe), Aristotle (384–322 BCE, pioneer of physics, biology, linguistics, the arts, and politics, teacher of Alexander the Great), Euclid (300 BCE, mathematician and physicist), and Archimedes (287–212 BCE, physicist, mathematician, and engineer). It is notable that science was studied along with other areas of human knowledge, including politics, the arts, logic, and ethics. The spirit of science therefore shared its roots with other areas of enquiry and knowledge, where rationality and logic gradually gained ground. The famous fresco *The School of Athens* by Raphael (Fig. 1.2) depicts leading figures of ancient Greek philosophy, encompassing what would later branch out into science and other areas of human knowledge.

As noticed earlier, Europe is not the only region contributing to the rise of the spirit of science. The spirit of enquiry and quest for knowledge was developed in many parts of the world. In Asia, in addition to contributions from China and the Indus region, the Arab civilization and Islam have also played significant parts, especially during the golden age from the 7th to the 15th century, in the development of mathematics, astronomy, chemistry, medicine, and engineering (Fig. 1.3). The first university, Al-Karaouine, was established in Morocco in 659 CE. The scientific method, consisting of asking questions, setting up hypotheses, doing experiments, and making analyses and conclusions, was used by Ibn al-Haytham, also known as Alhazen (965–1039), a Persian physicist, to study the properties of light. Muhammad Ibn Zakariya Razi (Alrazi, 854–925), a Persian physician and chemist, classified various materials and was the first to produce sulfuric acid. Ibn Sina (Avicenna, 980–1037), another Persian physician, is considered a forefather of modern medicine who conducted experimentation, observation, and critical reasoning in his scientific methods. More importantly, he is considered the crucial link between East and West, whose works brought back to the West from the Crusade Wars, along with the works of other Arabic scholars and Arabic translations of the works of ancient Greek philosophers, contributed to the age of Renaissance in the West.

**Figure 1.2** *The School of Athens* by Raphael, fresco painted between 1509 and 1511 and located at Apostolic Palace, Vatican City. It depicts many leading Greek philosophers, with Plato and Aristotle at the center. Public domain licensing.

**Figure 1.3** Science grew in many parts of the world, including the Middle East, India, and China. (a) Islamic science. Picture from Wikimedia Commons under Creative Commons license CC BY-SA 4.0 (https://creativecommons.org/licenses/by-sa/4.0/). (b) Chinese science (Ge Hong, official of Jin dynasty, 263–420, and alchemist). Picture from Wikimedia Commons. (c) Indian numerals, as now adopted universally (top) and in the Thai system (bottom).

The spirit of science grew rapidly in the age of Renaissance in Europe, the rebirth of a way of life with revival of the arts and sciences, beginning around the 14th century. At first it was rejuvenated by rediscovery of ancient Greek and Roman knowledge, in part brought over by Middle East scholars and European crusaders coming back from the wars to Europe. Later, there emerged new lines of knowledge from observation and calculation, which overthrew old beliefs, such as the formulation of the model of the universe by Nicolaus Copernicus (1473–1543), which placed the sun and not the earth as the center of the universe. Galileo Galilei (1564–1642) made a number of important physics and astronomical discoveries, such as the moons of Jupiter. His support of Copernicus put him in conflict with the Roman Catholic Church, an example of clashes between science and religious dogmatism. Johannes Kepler (1567–1630) discovered the laws of planetary motion and joined physics with astronomy. Sir Isaac Newton (1642–1727) discovered the laws of motion and the properties of light and co-invented calculus. Robert Boyle

(1627–1691) established many principles in chemistry and separated this science further from alchemy. In biology and medicine, William Harvey (1578 1657) discovered the circulation of blood, and later Robert Hooke (1635–1703) and Antonie van Leeuwenhoek (1632–1723) observed and described the cell and single-cell organisms, respectively, through microscopes they invented. These and other advances contributed to rapid development of modern science, based on observation and experimentation and not on unsupported speculation. The period beginning around the 16th century through to the end of the 18th century is called the period of the Scientific Revolution. The underpinning principles for this revolution are what can be construed as the scientific method, which philosophers like Francis Bacon (1561–1626) and René Descartes (1596–1650) helped establish. Briefly, Bacon developed the inductive method in science, where results of observation and experimentation are collected, analyzed, and built into an overall explanation. Descartes, on the other hand, advocated the deductive method, in which the scientist first formulates a hypothesis to explain phenomena and then seeks evidence to support or disprove the hypothesis. These principles of the scientific method have withstood the test of time up to the present day.

We can conclude that the spirit of science arose independently in many civilizations, stoked by the fires of curiosity, wonder, and exploratory urge. The spirit encompasses a long-lasting activity, which has flourished until today with increasing intensity. When we talk of a spirit, we mean something that does not die, as when we talk of the spirit of a person we believe that his or her character lives on after he or she has departed from the world. Likewise, when we talk of the spirit of a nation, we believe that the nation will live on and flourish, even though it may encounter difficulties from threats of war or domination by others. Science has a spirit, which is not only alive but has grown and evolved over the past millennia. As a body of knowledge about nature obtained through experiments and observation, science could be said to have originated back when humans discovered fire and other useful implements and they learned about nature from trial and error. It has gone through difficult periods in different parts of the world, such as Europe after the collapse of the Roman Empire and in other places with decay of their civilizations, but the spirit lived on in other places like the

Middle East and was rekindled in Europe with the help of contacts between them. From there, the spirit spread to other regions and was absorbed and reconciled with indigenous beliefs. Although some conflicts still remain, it is now truly a global spirit. It has now influenced the approaches to problems of large numbers of people from various countries and backgrounds, turning away from superstition, using reason and hands-on investigation to obtain knowledge, and using the knowledge and its products in their daily lives and in making their livelihoods.

A feature that ensures the survival and expansion of the spirit of science is its open-ended nature. This arises from the fact that any question asked will never gain a complete answer from a scientific investigation. Admittedly, the methods of science in observation and experimentation will yield some answer, but the answer obtained—if it is a good answer—will lead to more questions. Each level of answer is a measure of progress, encouraging the scientist to go on to the next level of questions and answers. For example, the question, what is the nature of genetic material? was answered in the middle of the past century—DNA. This was a big advance, leading to further questions on how DNA acts to transmit genetic characters to living things. These questions were subsequently answered through the study of other biological molecules involved, like ribonucleic acid (RNA) and various proteins, leading to both an understanding and applications of the knowledge gained to medicine, agriculture, and other areas. From the discovery that DNA is genetic material, the open-ended nature of science has led to many new areas of study, which is the mark of success of this line of investigation. Likewise, discovery of the nature of electricity as arising from the motion of electrons led to an understanding of conductors, semiconductors, and the science of electronics through further investigation. Investigation of the nature of atoms and their interactions led to expansion of modern chemistry and materials science, which helped in both explaining the world around us and making useful products ranging from drugs and cars to rockets. In addition to useful products, such investigation led to further questions on the nature of chemical reactions and the structure of materials. Through its open-ended and expansive nature, science has now involved so many areas that its content has been said to cover the "history of nearly everything" [2] (see Box 1.2).

**Box 1.2** Is Science Everything?

In the book *A Short History of Nearly Everything* [2], Bill Bryson said that science concerns "everything that has happened from the big bang to the rise of civilization—how we got from there, being nothing at all, to here, being us." He went on to describe a lot of things, including where we are in the universe, the nature of the stars, our world and what it contains down to the levels of subatomic components, how life began, how it evolved into a rich collection of interacting species, how "we are awfully lucky to be here," and how we should all try to make sure things do not end badly on this planet due to our recklessness. Altogether, the writer makes a strong argument that rather than being a dull subject, science is indeed an exciting one, which can provide an explanation or clues to an explanation of nearly everything that is interesting to an average person.

Looking at the list of subjects that science covers, some may be tempted to conclude that science is indeed everything, or at least almost everything. Not only does it cover the nature of everything around and inside us, from the stars to atoms and even smaller dimensions, it has now expanded to cover many aspects of social and human affairs formerly ascribed to nonscience. For example, the subject of politics has been viewed from the vantage of scientific investigation, giving rise to the term "political science." Others have tried to understand complex interactions of society in terms of biological evolution. Yet others have embraced an extreme view in believing that science constitutes the most authoritative accounts of human learning, the only real knowledge—a view termed "scientism." Historically, this belief can be traced back to the Scientific Revolution, starting from the 16th century, which helped launch the Enlightenment Movement two centuries later. The movement emphasized the importance of rational thought, religious tolerance, and individual liberty. It has a lasting influence, and many societal values of today were formed or elaborated during this time. The importance of science in creating and affirming knowledge was enhanced further in the school of thought called positivism, which states that authentic knowledge must be subject to analysis and verification, and the only valid knowledge is a scientific one.

Judging from the fact that science has been so successful in explaining natural phenomena, and the products of science are so pervasive in industry, transportation and communication, medicine and all modern human endeavors, and our everyday life in general, it is not surprising that the viewpoint that scientific knowledge is the only true knowledge has many followers. Yet with all the success of science in explaining natural phenomena, and in helping give birth to modern technologies with all their achievements, the view that science is all knowledge there is may be subject to scrutiny. The boundary for science is limited by the fact that it must deal with matters that are subject to observation and experimentation. Many concepts such as beauty, love, and hate appear to go beyond logical analysis and experimental verification and involve subjective judgment. Other areas such as morality and ethics may be subject partly to such analysis and verification, but they only provide data for discussion and debate that seem to go beyond the realm of science. Religion, which had many clashes with science in the past, some of which are not resolved even today, also involve areas of faith and beliefs that appear to be outside of the bounds of science. Other areas of human learning and endeavors, such as the arts, literature, and history, also appear to be largely separate from science, although of course scientific aspects can be found.

As Hamlet said to Horatio, "There are more things in heaven and earth, Horatio, than dreamt of in your philosophy." We may likewise need to heed this remark. However, we can still say that although science is not everything, it is surely nearly everything to do with the natural world and physical phenomena, especially when it comes to matters of imagination, investigation, and invention.

## 1.3 Knowledge Gained from the Spirit of Science

The spirit of science has grown stronger over the past few centuries, thanks to its success in enabling a deep understanding of nature and the application of this understanding to the development of various technologies, which we might call the sparks from the spirit. Unlike earlier civilizations, human societies all over the world are now well connected, in terms of both physical transport and information and communication technologies, resulting in a global pool of knowledge. Barring catastrophic wars or unforeseen global cataclysmic events, there is now no possibility that knowledge gained by humans over the course of history will be lost, like what happened in the past millennia. On the contrary, knowledge in different areas will tend to be merged together to give rise to "hybrid" sciences with new potentials for application or simply advances in understanding of nature. The following are examples of scientific knowledge, selected for illustration of basic principles and recent development in the integration of knowledge from various fields.

## 1.3.1 Physical Sciences

### 1.3.1.1 Earth science

Earth science is a composite science made up of components of geophysics (*geo* means earth), geology, geography, soil science, hydrology (study of water), meteorology (study of weather), chemistry, atmospheric science, and environmental biology. It studies the earth (Fig. 1.4), covering continental and ocean surfaces and the atmosphere that interacts with them, down into the depths of land and water masses, and up into the high reaches of the atmosphere, where the vast space lies beyond. It also covers the study of earth history, which can be subdivided into geologic timescales starting from the formation of the earth some 4.5 billion years ago to the present. Historically, this branch of science gained a big leap with Alfred Wegener's study of continental drift, which went from a period of general disbelief into established knowledge in the middle of the past century. This phenomenon concerns the movement of tectonic plates, or the plates on which the continents rest, as the primary mechanism that shapes the landscape. It is important to know the nature of our own earth, as, for example, the influence of magnetic and gravity fields on the pattern of the weather and water movements, seismic patterns that may result in earthquakes, the nature of land and oceanic sources of petroleum energy, and the nature of wind and solar sources. Earth science is pivotal to our understanding of climate change, due mainly to human activities such as those that produce earth-warming greenhouse gases, including carbon dioxide from fuel combustion. It has been proposed that we have entered a new geologic epoch, called the *Anthropocene epoch*, which sees big changes in the earth and the climate due to human activities. The proposed new epoch signifies a moment in world history where the risk to sustainability due to human activities is fast increasing.

**Figure 1.4** Earthrise. A picture of earth from space, taken by astronaut William Anders in 1968 during the Apollo 8 mission. Earth science comprises a study of the earth, both oceans and land and what lies under, and interaction with the atmosphere and what lies beyond. Public domain picture by NASA.

### 1.3.1.2 Space and fundamental physics

The nature of space and the universe was given a glimpse by early astronomers who helped start the Scientific Revolution. Newton gave an explanation of the movement of the stars in terms of his theory of gravity. Albert Einstein's theory of general relativity gave a new explanation of gravity in terms of distortion of space-time by masses. However, everyone assumed that the universe is static, until 1929, when Edwin Hubble discovered that the universe is expanding. This led to a logical conclusion that at some time in the past the universe must have started as a single, infinitesimally small point with a *big bang*. As it expanded and cooled off, it gave off microwave radiation, which would be detectable everywhere in the universe—a prediction that was later shown to be true. It has now been shown by other lines of evidence that the universe is expanding at an ever faster rate. What is the fate of the expanding universe? Will it go on expanding forever, or will it stop at some point in the future? Or will it even start collapsing on itself until we get a super massive black hole, and what will happen after that? No one knows the answer. Two factors will come into play, namely the existence of *dark matter* and the existence of *dark energy*. Their nature is not known, but it is now known that ordinary matter constitutes only some 5% of the total mass energy of the universe, while dark matter constitutes 27% and dark energy 68%. Until we know more about these two mysterious entities, the fate of our universe will still be unclear. Meanwhile, the existence of multiple universes, or multiverses, and the existence of more than four dimensions of space-time have been predicted by fundamental theories like string theory or M-Theory, which states that everything can be represented by a string or a membrane vibrating in multidimensional space (for more details, see, for example, Ref. [3]).

The span of physics ranges from the largest, like space and universe(s), to the smallest, like fundamental particles that constitute atoms, molecules, and all materials. Until a few decades ago, it was thought that the electrons, protons, and neutrons were the most basic constituents, but work on particle physics has changed all this. These particles are in fact composite, made of more fundamental particles. Protons and neutrons are heavy particles grouped as hadrons, which consist of quarks with various properties, and electrons and neutrinos (another elementary particle with no charge) are light particles grouped as leptons. The standard model of theoretical physics classifies all the fundamental particles and governs their interactions, which are basically of three types, namely electromagnetic, weak, and strong nuclear interactions. The discoveries of predicted particles, namely the top quark, the tau neutrino, and recently the Higgs boson, support the standard model. In spite of its success, the model is still not the theory of everything, since it still cannot fully explain the nature of gravity and the existence of dark energy and dark matter. The spirit of science should spur further development in these areas.

## 1.3.2 Chemistry and Materials Science

### 1.3.2.1 Structure of atoms and molecules

Modern concepts of atoms and molecules can be traced back to John Dalton in the early 19th century, who invoked the concept to explain why many elements react in small ratios of quantity, and to Amadeo Avogadro investigating the volumes of reacting gases and their products. The periodic table was later proposed by Dmitri Mendeleev to explain the periodic trends in the properties of the elements. Discoveries of the electron, the proton, and the neutron enabled the construction of the atomic structure with the protons and neutrons in the nucleus and the electrons revolving around it. This simple model had to be revised in light of advances in quantum mechanics, which states that matter exhibits wavelike behavior and can be located only by probability. Electrons are located in atomic orbitals, spaces where they are most likely to be, and chemical bonds are formed through interaction of the orbitals, which leads to a state of lower energy and more stability. These bonds are called covalent bonds, as distinct from ionic and other interactions between atoms, and are the links between atoms in a molecule. Various spectroscopic (using light or other radiations) and other techniques enabled the elucidation of molecular structures and the study of rates and mechanisms of reactions. Chemistry is often called the central science, closely related to many other fields. The hybrid science of biochemistry, for example, has led to the determination of complex structures of protein synthesis machinery and of the light-capturing mechanism in photosynthesis, contributing to the understanding of biology and energy transformation.

#### 1.3.2.2 Synthetic and natural product chemistry

Chemical substances are classified into organic and inorganic substances. The former are compounds that have carbon as a constituent, and comprise both those found in nature and those synthesized by humans, while the latter category covers all the rest. Because carbon can from four covalent bonds with itself or other atoms, the number of possible compounds is large, even with only a few atoms. The number of possible compounds with fewer than 70 atoms has been estimated to be something like $10^{60}$ [4]. This number shows the vastness of chemical space, or the space map containing all possible molecules. Techniques are now available for synthesis of many different compounds with desired molecular characteristics all together at the same time, rather than making them one by one.

Nature is a source of organic compounds, as indeed the term "organic" implies, but is now only a minor source since most organic compounds are now derived through synthetic methods. Organic compounds from nature, or natural products, are still good sources of, or provide starting points for, many current drugs and other useful materials [5]. Drugs or other compounds that are made starting from compounds derived from nature, such as from plants or microorganisms, followed by synthesis, are called semisynthetic compounds. There is also a trend in using biotechnological processes rather than conventional synthesis in making compounds: in this way the gene for accomplishing a reaction is cloned into an appropriate microorganism, which is then used as a "reactor" for producing desired compounds.

#### 1.3.2.3 Materials science

Materials science is a study of the structures and properties of materials, ranging from metals, ceramics, polymers, and biomaterials though composite materials, together with the methods for their fabrication. It is a hybrid science consisting of elements of chemistry, physics, biology, and engineering. The structures of materials are studied from the atomic level (dimension of $10^{-10}$ m) through nano- and microstructural levels ($10^{-9}$–$10^{-6}$ m). They can be studied through microscopic (using light, electrons, or atomic force), spectroscopic (using light and other radiations), and other methods. The bulk properties of materials include mechanical, electrical, thermal, chemical, optical, and magnetic properties. They depend on the types of atoms and molecules and interactions among them. Processing and engineering of materials are important in determining their bulk properties and performance. Bulk properties also depend on the sizes of particles that constitute the materials, as, for example, nanoparticles (dimension of 1–100 nm, or $10^{-9}$ m), which form various nanomaterials, such as surface coatings, textile, cosmetics, and medical formulations. Nanochemistry, a hybrid science of chemistry with physics and engineering, has led to the fabrication of nanomaterials such as carbon nanotubes, graphenes, and fullerenes with remarkable mechanical and electrical properties. Carbon nanotubes are used in applications such as energy storage, automotive parts, sporting goods, and electronic materials. There is some concern about the safety of nanoparticles, since they have dimensions that may allow penetration to various parts of the human body. These materials therefore are subject to extra safety management as a precaution.

## 1.3.3 Life and Medical Sciences

### 1.3.3.1 Molecular biology

The theory of evolution proposed by Charles Darwin and Alfred Wallace, together with Gregor Mendel's laws of inheritance, set the starting point in the identification of hereditary units, *genes*, which transmit genetic characters from parents to offspring. The momentous discovery by James Watson and Francis Crick in 1953 that such units are in the form of DNA connected biology with chemistry: genes and other components of life are chemicals, and their interactions are chemical interactions, albeit complex ones. The mechanisms of how genes are copied and transmitted, how repairs are made when mistakes occur, how genetic variations occur, and how genes interact with the environment are now within the realm of chemistry. Basically, the general flow of the process in the cell is from DNA to RNA and from RNA to proteins, with encoded information specified by the DNA. The combined total of all genes, called the genome, consisting of paired double strands, each with four basic components stringed together in unique sequences, is now known for many organisms. This includes the human genome with some three billion base pairs, the sequence of which was revealed at the beginning of the millennium. However, only a minor part of the genome specifies the structures of the proteins or control of their synthesis, while the functions of the rest, formerly believed to be "junk DNA," remains a mystery. At least some of this DNA is involved in the control of genetic processes, interaction with environmental factors, or development of complex organisms, the details of which remain to be found out.

The new area of synthetic biology is now engaged in making products of living cells, or living cells themselves, from synthetic genes (Fig. 1.5). A cell such as a bacterium can be taken over after its own genome has been replaced by a synthetic one. However, the complexity of interactions of genes with other components in the genome, other components in the cell, and other factors in the environment is still only beginning to be understood and should be the subject of intense research over the coming decades. Detailed molecular mechanisms of how organisms interact with one another also remain to be studied in detail. Development of knowledge in this area is fast; it is now possible to modify the genome using molecular tools such as enzymes that can be directed to precise positions to cut and modify genes or add new ones. The techniques pose a moral dilemma on to what extent we can modify our own genetic characters or even those of our offspring.

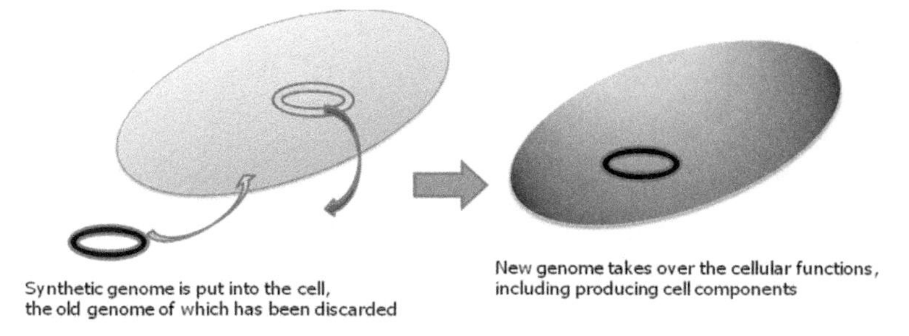


**Figure 1.5** Synthetic life can be made by substituting the natural genome containing all genes in a cell with a synthetic genome, which takes over all functions and produces all new cellular components.

### 1.3.3.2 Medical and veterinary sciences

From one viewpoint, medical and veterinary sciences are applied sciences, sparks of the spirit of basic science. However, much of them is basic in nature, delving into attempts to understand basic mechanisms of internal processes in humans and animals and their interactions with the environment, including disease pathogens, regardless of whether the findings will lead to cures or prevention of diseases. In the past century, classical sciences like anatomy (Fig. 1.6) and physiology have been transformed almost beyond recognition. New machineries and techniques in microscopy, such as electron microscopy and atomic force microscopy, have enabled visualization of microstructures down to the molecular level, for example, DNA and proteins. Noninvasive techniques such as magnetic resonance imaging have enabled detailed mapping and study of functions of the brain and other tissues in the living state.

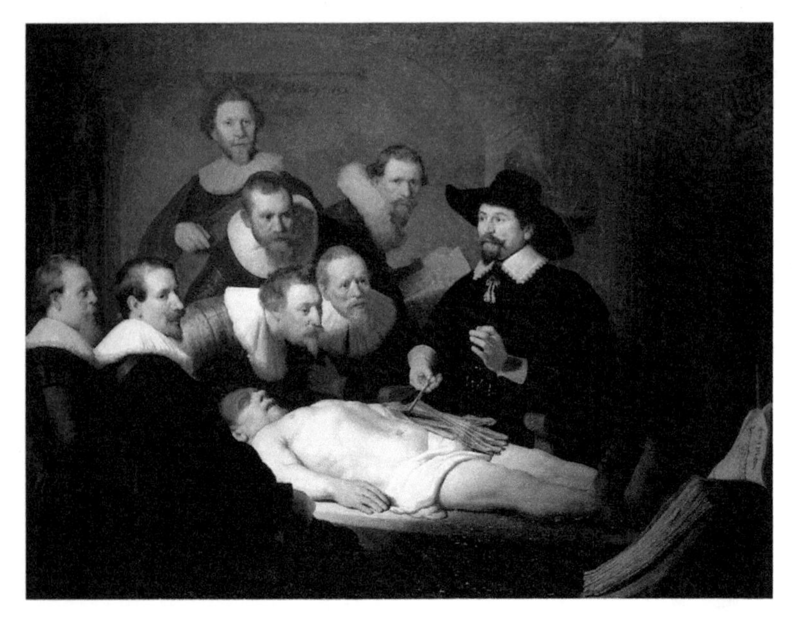

**Figure 1.6** Medical science has come far since *The Anatomy Lesson of Dr. Nicolaes Tulp,* an oil painting by Rembrandt, 1632, Royal Picture Gallery Mauritshaus, the Hague. Public domain licensing by copyright holder.

At the cellular level, we have learned how neurotransmitters, hormones, and foreign molecules like drugs exert their effects through interaction with receptors on the target cells, followed by transmission of signals into the cells. Many have successfully cloned tissues or even whole animals from stem cells, primordial cells that are unspecialized and can divide and grow taken from embryos or induced from adult sources. We have learned how cells control their division, including the roles of various genes, many of which—oncogenes, for example—lead to cancer upon being changed from a benign state by mutations. Other genes are used to fight off infections or build immunity of the organisms against invading pathogens. Two arms of immunity are active, antibody-mediated and cell-mediated, which work together in a concerted fashion such as in vaccinations that give effective defense against invading pathogens.

The threats of emerging diseases like bird flu, Ebola, or Zika make us realize that humans are vulnerable to many zoonotic diseases, or diseases endemic in animals. As for drugs against microbial infections, the discovery of penicillin by Alexander Fleming began the era of antibiotics, which was so successful at first that it was generally believed that the war against infectious diseases had been won. This proved to be a false dawn, as antibiotic resistance rendered most antibiotics ineffective only after a few years of usage. Searches for drugs that will not lead to drug resistance of the target pathogens are therefore of crucial importance. So is also the ability to predict efficacies of drugs in individuals, as it is known that drugs are not equally effective against different individuals, some of whom may even experience adverse effects of the drugs. The budding science of pharmacogenomics takes individual genomes into account in predicting the efficacy or likelihood of adverse events in drug usage. The genome is important not only for individuals but also for whole communities, where another budding science of genomic public health will help in determining policy in health care and disease prevention of communities.

#### 1.3.3.3 Plant sciences

The plant kingdom contains flowering plants, conifers, ferns, mosses, and algae. There are some 400,000 species of flowering plants alone, only a small fraction of which have been studied. It is important to study plants, not only because they provide us with so many basic things ranging from food, medicine, shelter, and in general much of the natural environment, but also because they form vital components of the biosphere, which contains all living species interacting with one another and with the environment. Plants are important in absorbing carbon dioxide from the atmosphere and giving back oxygen. They do so from their ability for photosynthesis, using light as the source of energy to produce carbohydrates from carbon dioxide and water. Carbohydrates are further converted to other components of the biomass, and plants are the source of foodstuffs for other organisms in the food chain. Photosynthesis comprises two parts, light and dark reactions. In brief, light reactions, occurring with the help of chlorophyll and complex components in the chloroplasts, use energy from light to prepare the energy source (adenosine triphosphate [ATP]) and the reducing power (nicotinamide adenine dinucleotide phosphate [NADPH]) for dark reactions, producing oxygen from water in the process. In dark reactions, also called the Calvin cycle, organic compounds are made from carbon dioxide using the energy source and the reducing power from the light reactions. An understanding of photosynthesis is not only academically important but also potentially useful for generating energy sources, including biofuels such as algae, and for engineering other systems for solar energy conversion based on this process.

The spirit of plant science has resulted in sparks with a major impact on agriculture, especially in the developing countries. Norman Borlaug and others, supported by philanthropic organizations and many governments, spearheaded plant-breeding efforts, resulting in improved varieties of wheat, rice, and other major crops. This movement ushered in what is called the Green Revolution around the middle of the past century. Selection was mostly done by examining the characters of crosses between varieties (called phenotypes), helped by genetic markers. The ability to manipulate genes a few decades later has enabled the construction of genetically modified organisms (GMOs) in a large variety of living things. Genetically modified (GM) plants, with a superior ability to produce grains and fruits, fight off pests, and withstand drought and other stresses, are now available commercially. However, some countries are still wary of using them, for fear of environmental and other hazards. Technologies are now available, not only for transplanting one or a few genes, but also for editing the whole genome so as to gain desired traits. Considering the fact that people in many developing countries are suffering from malnutrition and extreme poverty, the use of GM plants, or plants and animals modified by genome editing, in these countries is a major policy issue. The potential benefits and risks of the GM plants, and other GMOs, have to be studied carefully, not only in terms of science, but also for implications for trade, the environment, reduction of poverty and hunger, and sustainable development.

### 1.3.4 Human and Social Sciences

Human and social sciences use the scientific approach to ask and try to solve questions about ourselves as a human species, as individuals, and as people grouped together in societies. This area of study also includes human experiences in historical, cultural, and moral contexts and is therefore closely related to the fields of liberal studies and humanities that are traditionally separated from natural sciences. However, thinkers in the past, including David Hume, who coined the term "moral science," included these areas as legitimate for scientific studies based on empirical observation. Adam Smith also conceived of economics as a moral science. Modern studies of human and social sciences include the fields of anthropology, psychology, population science, and ethics. Sociology, economics, history, and politics can also be included, although these fields are now large by themselves and concern areas that are outside the traditional boundaries of science (see Box 1.2). A main point of contention in social and human sciences is that it involves not only objective phenomena that can be observed and studied through the normal scientific methods but also human perception and attitude, which are subjective in nature. While subjective judgments without rational basis are outside the realm of science, many argue that more consideration should be given to the role of human consciousness, intuitive senses, and holistic aspects of the social and human problems.

Human and social sciences play a prominent role in our quest for development and sustainability, which should be considered the main sparks from the spirit of science, in addition to scientific and engineering innovations. Indeed, social innovations such as social networking, social enterprises, and philanthropy are intricately linked with development in social and human sciences, and not just with the hard sciences and engineering like information and communication technologies. We need to understand human nature and human society more in order to try to build a sustainable world where poverty, hunger, diseases, inequality, and human conflicts are greatly reduced, where resources are more equitably distributed, and where consumption and production patterns are more sustainable, with minimum adverse effects on the climate and the environment. Social and human sciences should be an integral part in these momentous efforts.

## 1.4 Turning the Spirit into Action

How does one turn the spirit of science into action? In short, how do you do science? We saw earlier that the spirit of science is driven mainly by curiosity, a sense of wonder, and willingness to explore various avenues to find answers, or at least clues to answers. Indeed, we do try to find answers to questions that arise in everyday life all the time. When we feel unwell, we ask ourselves whether we have eaten something wrong, are starting to catch a cold, or simply are too weary from our workload. We then try to recall whether there could be something in the food or water, whether we stood in the rain, or whether we did not take enough rest. We can further decide on the course of action, whether to consult a doctor, look up for information on the Internet, or simply take care of ourselves according to our assessment. When scientists pursue their work, they also ask questions and try to answer them. The difference is that in professional scientific work, a question has to be asked with clarity, and it has to be a question with potential answers containing real information, not just rhetoric or something that can be answered only generally without real information. If the scientist is working on a research problem, then the potential answer should be a novel one. The important thing here is that the answer should be obtained through observation or designed experiments. If it is a question concerning theories, the answer can be a theoretical one, preferably backed by mathematical argument. The answer must generally obey the accepted principles of science, generally said to constitute the paradigm of normal science, and should preferably lead to further questions that advance the state of knowledge further.

This rather abstract explanation can be made clearer with a real example. Take the case of the illness we talked about earlier. In this case, suppose you decide to consult a doctor, and suppose it turns out to be a complaint that fits a case of influenza. In a normal case, the doctor may prescribe some painkiller and some decongestant. However, in cases where the doctor has some reason to think that it may not be a normal case of the flu, such as the presence of unusual symptoms, the doctor may turn to the scientist to investigate further. The scientist can find out possible a causative pathogenic organism, which could be an unfamiliar virus. The scientist then needs to find out whether the virus is known or whether it is a new variety. If it is a new variety, it should be characterized and reported to the medical and public health communities, in terms of the genetic characters and properties, medical symptoms associated with human infection, and caution regarding its contagiousness. The scientific community will then home in on the investigation of the nature of the virus and its implications on the health of individuals and public health. In time, studies will be made to find ways for diagnosis, prevention, and therapy of the illness caused by the virus.

It is important to note that in the work of the scientist, the problem-solving principle devised by William of Ockham (or Occam, 1287–1347), known as Occam's razor, should be used. In brief, the principle states that "among competing hypotheses, the one with the fewest assumptions should be selected." In other words, choose the simplest explanation over others. Furthermore, the explanation should lead to predictions that can be tested. If the predictions are proved to be wrong, the explanation is invalid, and another explanation must be found. Coming back to the example of investigating the illness that fits description of a bout of the flu, one should first make the hypothesis that it is caused by an infectious agent, rather than being caused by an unusual psychological state or a curse by someone. In all three cases, there are assumptions involved, but the first case with an infectious agent as the cause is the most straightforward and leads to the prediction that it must be identifiable and studied further to show that it does indeed cause the illness. In contrast, in the other two cases, there are assumptions that cannot be substantiated by evidence and do not lead to testable predictions. In the history of science, there are many instances where Occam's razor, backed by experiments, proved to be the best tool for progress. One case in point is the former belief that living things have a vital force (or vital spark) that distinguishes them from nonliving things. The fact that organic molecules and biomolecules, including synthetic genomes, and even whole life forms, can be made artificially renders the concept unnecessary and can be cut away by Occam's razor.

Although most of science proceeds by investigating or observing phenomena and invoking explanations through accepted principles and theories, the spirit of science also demands skepticism and rebelliousness. This is an important feature, since science proceeds not only by accumulating information that supports existing principles and theories but also by replacing them with more elegant ones. For example, Newton's laws of motion, although satisfactory for an explanation of general phenomena, need to be replaced by Einstein's theory of special relativity when the velocity of an object approaches that of light. Until a few years ago, it was thought that infectious diseases could be transmitted only by agents with genes, until it was shown that in some cases proteins can be the infectious agent, most notoriously the mad-cow prion disease. These big changes in scientific principles and theories are called paradigm changes and are important for the development of science so that it can approach the best explanation of nature. They show the open-ended nature of science, always in a state of evolution, and nothing is considered sacred or untouchable.

# Chapter 2

# The Spirit Has Sparks

*A little spark is followed by a great flame.*

—Dante Alighieri, *Divine Comedy*, 1472
(H. W. Longfellow's translation)

The spirit of science, combined with human ingenuity in other areas, including tool making, art and design, entrepreneurship, and commercialization, leads to sparks in countless human affairs. They often start as little sparks, which, if the conditions are right, are followed by big flames. Sparks can be sparks of ideas in science itself, leading to a better understanding of nature and applications to human society. Other sparks lead to innovations that enhance the fields of agriculture, industry, health, education, trade, and services. They are embedded in everyday life, in the forms of things we use, how we work, communicate, travel, relax, and live in general. Innovations in the form of products and processes can be counted as the first sparks from the spirit of science. Together with good policy and good governance, they lead to subsequent sparks of development. These usually come from collective actions of individuals, groups, or societies over many years, with contributions from people other than scientists and technologists, including financial managers, industrialists, and public workers. They may have unforeseen and unwanted side effects, or they may be superseded by newer and better developments. Good governance

*Sparks from the Spirit: From Science to Innovation, Development, and Sustainability*
Yongyuth Yuthavong

ISBN 978-981-4774-57-4 (Hardcover), 978-1-315-14599-0 (eBook)
www.panstanford.com

and proper management, including mitigation and prevention of the undesirable effects, can lead to sustainable development based on science and its innovations.

## 2.1 Sparks from the Spirit and Their Effects

Science has the spirit that inspires and energizes people to find answers about various things in nature, ranging from the stars to the atoms and from our own life processes to nuclear reactions. That is only one part of the picture. The spirit of science has sparks, in the sense that the knowledge acquired leads on to applications, both in terms of techniques and skills to accomplish various tasks and in terms of machines and products derived from such skills and techniques. We call these skills and techniques, including the knowledge embedded in them and in the machines and products, collectively as technology. Modern technologies such as information technology (IT) and biomedical technology are mostly based in science. Technologies that had their origins long ago, like agricultural technology, irrigation, and building technology, grew in parallel with science in the history of human development. Many of the skills and techniques in these long-developed technologies were derived from experience and trial-and-error procedures, not necessarily as derivatives of science. However, in the present age, all technologies, including these developed long ago, rely mainly on scientific information and tools derived from it. For example, old civilizations may have been able to manage their water resources through the building of reservoirs and canals from indigenous knowledge, but we are now much more capable of doing so using knowledge from a variety of scientific fields, including hydrology (science of movement, distribution, and quality of water), earth and environmental science, and construction engineering. Old civilizations were almost helpless—indeed many were wiped out—when deadly infectious diseases like plague and cholera struck. Nowadays, with knowledge from life sciences, including microbiology, we can prevent and control the spread of such infectious diseases through vaccines, drugs, and hygiene, all of which are technological derivatives of the spirit of scientific investigation. A large-scale spread of plague, cholera, and other deadly diseases is now unlikely, although a few cases may occur periodically. Even new, emerging diseases like Zika, bird flu, or severe acute respiratory syndrome (SARS) are well contained and suppressed in relatively short times, not only through medical sciences, but also through vigilance, containment strategies, and fast reporting with the help of IT. Science-based technology is so important and tightly woven with science itself that the two are often referred to together as science and technology.

Because of the importance of technology in trade, industry, and everyday life, the sparks from the spirit of science demand much more attention from society than science itself. There is a continuous spectrum, from basic science through applied science to technology. Applied science is different from basic science in that while the latter is done with curiosity as the main driving force, the former is done with an aim toward application and is very closely linked with technology. In view of the fact that societies depend critically on the success of their technology, the allocation of resources and personnel is, not surprisingly, mostly for technology and applied science, while basic science has a much smaller share. History teaches us, however, not to ignore basic science, since it is the wellspring of ideas and knowledge that lead to major breakthroughs. Thermodynamics, the science of heat, led to the development of the steam engine and is the basis for the energy technology of today. It is possible that nuclear fusion energy, stemming from basic research in physics, will be a major source of energy for tomorrow. Biotechnology industries of today started from basic discoveries in molecular biology only a few years ago. Electronics and computer technology started off as offshoots of physics and materials science. It is the spirit of science that leads to the sparks, including major advances in technology, with further repercussions for sustainable development.

The sparks from the spirit of science may start very slowly, almost unnoticed at first, and then grow gradually. They may also develop independently in different locations according to different needs and environments. An example is improvement in crop rotation, or the planned sequence of growing different crops from one season to the next, which has long been practiced by farmers the world over. This practice was shown by the experience of farmers and through scientific investigations to lead to improved crop yields, reduced soil erosion, and better pest control. Farmers in each geographic area have to find the best combination of different crops, which also have to be chosen with consideration of the market. Successful farmers are the ones with the spirit of science, and their experience contributes significantly to our knowledge about the importance of having a variety of plants in a location, called polyculture, for the environment and agriculture. Agricultural scientists can work further to develop innovations in crop rotation with regard to various factors such as the use of new crop varieties, farm size, need for water and fertilizer, change of climate, and other factors.

In some cases, the sparks may be quite substantial, resulting in, say, a new product or a new process. The various inventions of Thomas Alva Edison, including the light bulb, the camera, and the phonograph, come to mind. The discovery of antibiotics by Alexander Fleming, followed by success in industrial production, changed the course of medical history. More recently, discoveries and inventions have transformed the way we live substantially, so much so that a person who lived a hundred years ago would have great difficulties in understanding and using the tools that we now use daily, including the computer, cell phones, household appliances, and transport vehicles. The discoveries and inventions that enable us to lead our lives with higher quality, protect or cure us from illness, create new markets, and increase trade and industry can collectively be called innovations. Innovations have positive effects on individuals, communities, governments, and societies in general, although in many cases they are disruptive in introducing changes that threaten the "business as usual" life. If managed properly, innovations bring wealth, security, enjoyment, and comfort to our lives. They form a basis for the development of societies, ranging from those of very poor countries to wealthy ones. Moreover, the development need not be short lived, like those earlier civilizations that were later torn down by wars and pestilence.

Often, what was originally thought of as belonging to the realm of basic science without prospects of direct applications turns out to be very useful when the time comes. Einstein's theories of special and general relativity were long thought to be parts of basic physics without practical benefit until the present age. In the development of the Global Positioning System (GPS), for example, there is a need to have precise information on time at various positions, which according to these theories would depend on the relative speeds of movement of satellites and the gravitational pull of the earth. Only with the corrections enabled by these theories could GPS technology advance to the present state of precision. Sparks may, therefore, arise when the conditions are ready, and there should be room for scientists to pursue problems in basic science of their interest. The scientists, in return, should be on the alert for potential applications of their basic work.

Figure 2.1 summarizes the relationship between science and technology and their output as discovery and invention. The sparks can be produced as ideas for knowledge or for innovation as products and processes with benefits to society. As we will see in this book, we now have the tools to bring us not only short-lived development but also development that can be sustainable, the outcome of which has a lasting effect in the long run. These tools, however, are not just those from science and technology but also those from other spheres of human activities, including politics, diplomacy, and social and cultural studies. The innovations from these other spheres of human activities are called social innovations, and we need them together with technological innovations in order to achieve development, in particular sustainable development.

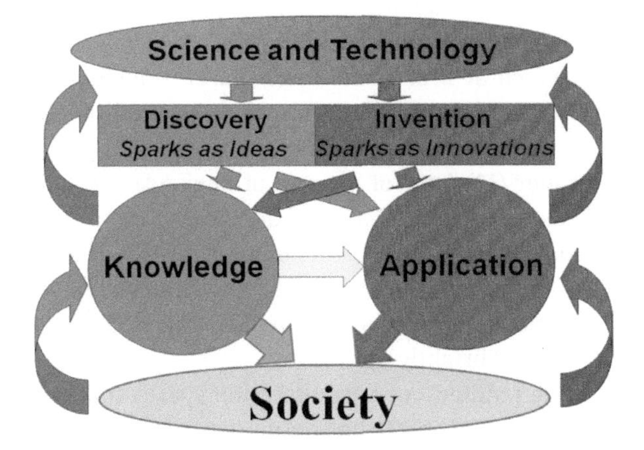


**Figure 2.1** A simplified diagram showing the relationship of science and technology with discovery and invention, leading to knowledge and application for the benefit of society. Generally speaking, science leads to discovery and knowledge, while technology leads to invention and application in various productions and processes. Feedback arrows indicate that the domains are interactive, as, for example, the demand of society can have positive effects on generation of knowledge and application. Sparks can be formed as ideas or innovations, leading to further societal development.

## 2.2 Little Sparks, Great Flames

In Dante's "Paradiso", the last part of his *Divine Comedy*, in which the poet described his journey to the heavens, he hoped that his undertaking—a tiny spark—would be followed by greater acts of others. In real life there are many examples where small starts lead to big actions with great results. These examples can be found in acts that come from strong conviction. The work of Wilbur and Orville Wright in inventing the first successful airplane and making the first powered and controlled human flight in 1903 is a celebrated example of a little spark of genius and perseverance, which led to our present great age of flying. Looking back, it is easy to forget that the spark, little though it was at the beginning, did not come out of nowhere. The Wright brothers spent many years studying earlier works on gliding and flight, ranging from the drawings of Leonardo da Vinci to models of other flight enthusiasts like Otto Lilienthal (1848–1896), and making designs for wings, engines, and control mechanisms that were superior to the earlier pioneers. Therefore, even this little spark had a long-burning fire to start it off.

Let us consider another little spark that led to big flames. Michael Faraday (1791–1867) was the pioneer of research on electricity and magnetism, which at first formed only parts of investigation into basic physics. There was a story that when he tried to explain the importance of his discoveries to William Gladstone, then chancellor of the exchequer (Minister of Finance of England) and later prime minister, Gladstone asked, "But, after all, what use is it?" "Why, sir," replied Faraday, "there is every probability that you will soon be able to tax it." The accuracy of this story is disputed, but it is clear that the spark from Faraday's early work was indeed followed by big flames, which have burned brightly until today.

It should be pointed out here that little sparks that lead to big flames are not confined only to inventions or other outputs of science. Indeed there are many sparks from other areas of human activities that lead to big flames of great benefits to human society. The work of Florence Nightingale in helping soldiers during the Crimean War, which led to the beginning of the nursing profession, and the work of Henry Dunant, who organized the civilian population to help the wounded in the Battle of Solferino, leading to the creation of the Red Cross, are examples of sparks initiated by lone individuals with conviction and courage, which won the acclaim and participation of the people at large. In the area of cooperatives (organizations that are owned and run jointly by their members), Benjamin Franklin in the United States and the Rochdale Pioneers in England were the early founders who started the sparks that led to big flames of cooperative movements in societies the world over. For our purpose, we should just note that these sparks and those from the spirit of science share the same features, that is, being difficult to ignite or carry at first, but owing to the benefits they ensue and also in many cases due to lucky circumstances, the sparks went on to become big flames with a large impact on human welfare and development.

We see that the spirit of science produces sparks that can generate useful things, big or small. We recognize many types of sparks, from innovations to development and sustainability. Furthermore, the spirit of science can also be a model for spirit in other spheres of activity, like education, law, management, and problem-solving tasks in general. Some sparks may come entirely by accident, unplanned events—called serendipity—that trigger discoveries or inventions (see Box 2.1). Such serendipitous events, which may otherwise pass by unnoticed, are recognized by the prepared minds as clues to discoveries or inventions. Other sparks may take long and diligent hard work before yielding expected results as discoveries or inventions. While serendipitous events are more dramatic and often used as examples, it is more common to be engaged in long and patient hard work before the sparks can be ignited into something really significant and useful.

**Box 2.1** Tales of Serendipity

Serendipity (pronounced *ser-en-dip-i-ty*) must be one of the most difficult of all English words, both to pronounce and to understand without extended explanation. Indeed, the word is not originally English but was coined by Horace Walpole only in 1754 from a Persian fairy tale, "The Three Princes of Serendip." Serendip is an old name for Sri Lanka, where the three princes came from. The princes were always making accidental discoveries of things they did not expect or plan, but were able to use acute reasoning for an explanation. The word has come to be widely used to mark a discovery or invention that came by accident.

Perhaps the first person who found out how to light a fire back in the dawn of human evolution did so by accident, but we will never know. The first recorded celebrated case of serendipity was from a time long before the word was invented. King Hiero II of Syracuse reportedly asked Archimedes to assess the purity of his golden crown. This would mean finding out the density of the crown, which would be known if the weight and the volume were known. While the weight was easily measured, how could the volume of an irregular object like the crown be known? While taking a bath in a tub, Archimedes noted that the level of the water rose as he went in. This was the solution he was looking for—the volume of an object can be measured from the volume of water it displaces. The story goes that Archimedes was so excited about this discovery that he ran out of the bath naked, shouting "Eureka! Eureka!" (Fig. 2.2).

**Figure 2.2** The first recorded case of serendipity was a naked one—the "Eureka!" moment of Archimedes.

The discovery of the antibiotic penicillin in 1928 by Fleming, a Scottish biologist and pharmacologist, is a modern story of serendipity. He grew the bacterium *Staphylococcus* (causing boils, sore throats, etc.) on petri dishes and went on holiday without cleaning them up. Coming back, he noted that some dishes had mold growth, with clear bacteria-free zones around the mold, as though it was producing something that killed off the bacteria. The mold was later identified as *Penicillium*, and its "juice" from a culture of the mold was found to kill a lot of other bacteria. Subsequent work by Oxford researchers resulted in purification and identification of the structure of the active ingredient, which was later shown to be active in clinical trials, and its production by fermentation was scaled up in time to help save lives of many wounded soldiers in the Second World War. This tale of serendipity shows that an important discovery may occur just from observing uncleaned petri dishes so long as the researcher notices and uses reason to follow up on the observation. Even then, it took more than a decade to turn penicillin from a curious discovery to a life-saving drug by way of cooperation of many researchers who took up the issue seriously.

Percy Spencer, a man with little formal education, was working on building magnetrons (high-power radiation-producing vacuum tubes) for radar sets for the Raytheon company in 1939. He noticed that the candy bar he had in his pocket melted when he was standing in front of the machine. Although he was not the first to notice something like this, he was curious and began investigating into the matter. He and his colleagues found that the microwave radiation from the machine caused the heating, and an improved machine was made to direct the radiation to substances, such as foods, in a specially designed container. The first commercial machine was much bigger, heavier, and more expensive than today's microwave ovens, which are the products of many improvements over the years. This account teaches us not to overlook something unusual, such as chocolate melting when it should not, and to investigate possible causes. It also tells us that such an important tool, with potential for use by every household, would over time be developed to meet market demand once the idea for the initial invention had been conceived.

The last tale here is a tale of double serendipity. The first one was when Spencer Silver was trying to make a superstrong adhesive while working at the 3M company in 1968. Instead, he got a new polymer adhesive that would rearrange itself into small spheres on a surface rather than covering it evenly, hence behaving as a weak adhesive that would stick to surfaces on pressing but could be removed and reattached easily. At first no one found good use for this until the second serendipitous moment. Art Fry, an amateur singer also working at 3M, had an idea to use this adhesive to make removable stick slips for music note books he was using in giving concerts, instead of normal bookmarks that kept falling off. He suggested using the adhesive to make commercial removable stick notes for papers and books. Even then it was doubtful as a marketable product, and the company decided to distribute the product free for a year. Success came as more than 90% of the users indicated that they would buy the product, which was marketed later as Post-it notes. The moral of this story is that a useful invention may have to wait until someone figures out how to use it and the public at large finally appreciate its benefits.

There are many other accounts of discovery by serendipity, ranging from Teflon to Velcro to potato chips. Opportunities present themselves to everyone over a life time, but most are missed or not followed up. As Louis Pasteur, the famous chemist and inventor, said, "In the fields of observation chance favors only the prepared mind." This was true for Archimedes then, and is true for everyone now.

## 2.3 Sparks as Ideas

We first examine the nature of sparks that are offshoots of scientific investigations. These are new ideas generated in the light of scientific discoveries. Scientific discoveries are the major aims of scientists and often have big knock-on effects. In addition to contributing to the stock of knowledge, discoveries lead to ideas on possible benefits or implications for further development, either of the scientific knowledge itself or of broader issues. In what follows we shall examine three cases of ideas that are offshoots of discoveries, the first one of which was rather clear right from the beginning, the second developed from the main discovery, and the third still has unclear implications.

Wilhelm Roentgen, a German professor of physics, discovered X-rays in 1895, for which he was awarded the first Nobel Prize in Physics in 1901. He observed emitted radiation that penetrated objects as he was studying the passage of current through an evacuated tube. Although others may have noted the effect of the X-ray before, Roentgen was the first to study it systematically and to demonstrate its nature, especially in being able to penetrate objects that are not heavy metals. He clearly appreciated the wide implications of his discovery, as he produced the image of his wife's hand and analyzed the variable transparencies of the bones, flesh, and the wedding ring. This was the spark that started the field of X-ray application, which has stayed in use until today. Not only have X-rays revolutionized the field of medical examination, but they later led to wide-ranging applications, such as X-ray spectroscopy to study stars and other objects in the universe and X-ray diffraction to study the structure of atoms and molecules.

James Watson and Francis Crick discovered the structure of deoxyribonucleic acid (DNA), the molecule that transmits genetic messages, as a double helix of paired complementary strands, each of which carries the genetic code as a linear sequence of four "letters" (chemical bases A, T, C, and G). They noted the implication of their momentous discovery in the last sentence of their short paper in 1953 [6]: "It has not escaped our notice that the specific pairing we have postulated immediately suggests a possible copying mechanism for the genetic material." Their discovery later spawned ideas that led to many related discoveries and development, such as how the genetic material is copied and the information transferred to ribonucleic acid (RNA) and subsequently proteins, which are major components of living beings. Now, some seven decades later, genomes, or the full sequences of genetic letters of individuals, are widely available. Many of the sparks resulting from their initial discoveries arose later, such as technology for amplifying and sequencing small amounts of DNA, which makes it possible to identify criminals, study ancient remains of animals and plants, and diagnose genetic conditions of unborn babies, to name but a few examples. Indeed, whole new subjects of genomic medicine and genomic public health are springing up to take us into a new era where individuals and the public will obtain benefit from genetic diagnosis, gene therapy (therapy of diseases by modifying the genes of the patient), and other genetic technologies.

The last example of sparks as ideas delves into an as yet unsettled scientific hypothesis that would have enormous implications for the future of humankind and of living species on the earth in general. This example has been chosen partly because of its relevance to the issues of sustainability, which we will discuss later. The Gaia (read *gay-ah*) hypothesis was proposed by a chemist James Lovelock in 1979, based partly on a similar hypothesis proposed 200 years earlier by a Scottish geologist James Hutton that the earth (including all living beings and everything in the environment) acts as a single living system that can regulate itself [7]. The name "Gaia" comes from the Greek goddess of the earth, who pulled the living world together out of chaos. Just as a whole living being has living cells and organic and inorganic matter as its interacting components, so does the whole earth, with various species interacting with one another and with the environment in a self-regulating manner. Self-regulation was shown in a mathematical model ("daisy world" model) of a simplified world with only two species, black and white daisies, the former absorbing sunlight, resulting in warming, and the latter scattering sunlight, resulting in cooling. In this model, a scenario where sunlight becomes increasingly strong does not significantly change the surface temperature of the daisy world, since there is responsive change in the ratio of the two species, resulting in temperature regulation. The Gaia hypothesis received support from a number of people. However, it is still not generally accepted by the scientific community, some members of which regard this only as an analogy with individual living beings. This last example has been chosen for discussion here because there are many implications for human development and environmental sustainability. We now realize that as a consequence of human activities in energy usage, agricultural practice, and our ways of life in general, the world is undergoing significant changes in climate and the environment. Will the world be able to regulate itself and come to a new equilibrium in which human activities can be accommodated? Or will the human species be analogous to an invading pathogen, pushing the world to a catastrophic situation in which the human species itself is in peril together with other components of the living earth? These are questions we need to answer, no matter whether the Gaia hypothesis will be accepted in the scientific domain or remain just as an analogy. It has given rise to many sparks of ideas for further development.

## 2.4 Sparks as Innovations

We next examine the sparks that go beyond ideas into real products and processes that are called innovations. As noted earlier, discoveries and inventions that lead to a better quality of life, create new markets, and increase trade and industry can collectively be called innovations. Scientific discoveries alone are not innovations in the sense that they do not lead instantly to useful products and processes. However, they lead to new ideas, as we saw in the earlier section, and ideas can lead to useful innovations. Many useful innovations come from ideas not necessarily embedded only in scientific but also from practical considerations. One most important criterion for successful innovations is that they must be amenable for use by people, either through the market or through general use as public goods. Here we examine three examples of sparks as innovations.

Our first example goes back more than two centuries ago. James Watt (1736–1819) is credited with the invention of the steam engine. In fact, steam engines had already been invented even before Watt was born. One known as the Newcomen engine was used to pump water out of mines in England. The engine was a cylinder with a piston, and after steam was let in it was condensed by cooling, creating a vacuum that forced down the piston to do the pumping work. However, these engines were quite inefficient, since the cylinder and the piston had to be cooled to condense the steam, hence wasting energy through the heating–cooling cycle. Working as an instrument maker in University of Glasgow, Watt saw that he could improve the engine's efficiency by having a condensing unit separate from, but connected to, the cylinder, which therefore did not need to be cooled and reheated repeatedly. He also invented an engine in which steam could be introduced to both sides of the piston, resulting in a double-acting engine, and a mechanism for moving the piston up and down from a rotating wheel. After engineering improvements to the design, the Watt engine became very successful through these innovations, but its success was also due considerably to his successful collaboration with Matthew Boulton, who brought the manufacture and marketing of the engine to the industrial scale. It was soon applied to other tasks than water pumping, such as in ironworks and cotton mills, and later to drive train locomotives, ships, and many other power machines. The horsepower was defined by Watt as a unit of power; his name "Watt" was later adopted as a universal unit of power.

Fast-forward to the present age, in which a notable innovation is a spark from computer and IT. While working at the European Centre for Nuclear Research (CERN) and for some time in a private company in the 1980s, Tim Berners-Lee conceived of a concept for communicating among computers through a system later called the World Wide Web. It was based on the use of hypertext, text that includes letters, graphics, video, and sound, linked to other texts in the network through the Internet. A similar concept was earlier explored by a noted engineer and science administrator Vannevar Bush, who conceived of a *memex*, a microfilm machine that can link and integrate information from various sources into a giant encyclopedia. The advent of the high-speed Internet made possible the linkage as developed in the World Wide Web, which is now a major mode of communication, including transmission of information for business, education, and other purposes. It led to development in e-business, e-information, e-learning, and social networking such as Facebook and Twitter. It can therefore be said to be an innovation that truly changed how humans communicate with one another. Berners-Lee is a strong advocate of using the Web as a tool for a democratic system of accessing information and expressing opinions but with protection of privacy. He has refused to gain commercial benefits from his innovations.

Our third example is the discovery and development of effective antimalarial drugs by the Chinese cooperative research group, of which Youyou Tu was a member and to whom the Nobel Prize was awarded in 2015. The group comprised some 500 scientists from 60 institutes, participating in a secret task during the 1960s and 1970s of finding drugs from traditional medicine information, which would help China strategically in conflicts with the West. They followed the instruction from Ge Hong (284–346) (see Fig. 1.1b, Chapter 1) in obtaining extracts from a plant called *qinghao*, followed by a chemical and pharmacological investigation of the active principles. An extract found to cure mice of malaria contains a compound called artemisinin, which was later shown to be very active against human malaria as well. The structure of the compound was elucidated, and turned out to be a relatively simple one with a peroxide chemical group essential for its activity. Derivatives were made, some of which were shown to be even more active. Active compounds went through clinical trials and were manufactured by government pharmaceutical laboratories for malaria treatment. The compounds gained acceptance worldwide as some of the most active, fast-acting antimalarials and are now used in combination with other antimalarials for standard malaria therapy.

These different cases of sparks of innovation show some common characteristics. They are based on scientific principles but require more than ideas from such principles, including hard practical work and problem-solving capabilities found in good engineers and professional practitioners. Furthermore, they need actors other than scientists and engineers to move from the stages of discoveries and ideas to actual manufacture and implementation on a large scale—the manufacturer and business partner for the steam engine, the information network business for the Web, and the Chinese government and later the pharmaceutical industry for the antimalarials. Last in the steps to successful innovations, they must be widely adopted by users, including the public and selected sectors such as industry or medicine, due to the benefits they bring in either as new capabilities or as the ability to solve existing problems.

## 2.5 Management of Sparks: Reaping and Sharing Benefits

In the past, innovation used to be the fruits of talented people who had the sparks for inventing useful products or processes. As industry developed, progressive firms undertook research and development (R&D) as an important part of their strategies, and the competition between them is often won by the party that has superior technologies and innovations. Many universities also highlight the role of innovation together with research. Management of innovation helps firms and universities achieve best results in terms of encouraging new ideas, supporting inventors in the development of new products and processes, and bringing them to the market. In the standard practice of industry, innovations may be patented or registered otherwise as intellectual property. Some innovations are protected by other means, such as trade secrets and copyrights. It is only fair that the innovators or owners of the innovations benefit appropriately from their work, and these intellectual properties are protected by licensing or permission from the owners. More importantly, protection of intellectual properties is a mechanism to ensure that the discoveries and inventions can be developed for the benefit of the consumers, since they will be amenable to production and marketing by industry without fear of theft of ideas and techniques and imitation of the inventions. The conventional means for protection of intellectual properties like patenting work well for large industrial companies, with researchers and inventors who start the process of innovation, through to development of the products and processes, including scaling up of production, to mass production and marketing of the products.

There have been recent shifts in the concept of innovation, especially in new industries namely information-based (e.g., smart phones, telecommunication) and biobased (e.g., new drugs, biomaterials) industries. The conventional mode of innovation might be called closed innovation, where the developers of products and processes rely on the advantage of being the first to market and excluding competitors from the field. However, innovators in the new industries tend to be young entrepreneurs who form their own companies and offer radically new product lines that do not fit with the policy of large, risk-averse conventional companies. People like Bill Gates (Microsoft) and Steve Jobs (Apple) in computer technology and Mark Zuckerberg (Facebook) and Sergey Brin and Larry Page (Google) in social media offer examples of successful new innovators. Many innovators of the new industries are researchers from universities, who might set up or be partners in small and specialized biotechnology or software development companies. They are mobile and unlikely to be attracted to a lifelong career in big companies. The availability of venture capital in the money market, capital provided by investors to fund early development of new technologies or new-growth companies, gives incentives to risk-taking innovators and entrepreneurs. Many big, forward-looking companies also look to outside sources of technology in the form of intellectual properties for which they can obtain a license or through acquiring small companies. The innovation process is not just confined to a company but is achieved by a network of partners, including big and small companies and individual innovators. In contrast to the traditional style of closed innovation, a new process called open innovation is being increasingly adopted. Many innovators, especially those from universities or research institutes, argue that innovation should be public goods and call for open innovations. Often, a combined strategy is used. Open innovation is practiced in the early stages of generating ideas and directions so as to have sharing of knowledge at the generic stage, while closed innovation is adopted when products or processes with commercial interests are being developed.

## 2.6 Sparks as Development

Development is defined as the act or process of growing or causing something to grow or become larger or more advanced. The sparks from the spirit of science that we have already encountered as ideas and innovations can lead to development, such as what we already saw in the use of DNA technology, which led to the development of medical and health industries, and the use of the World Wide Web, which led to the development of new business, new ways of learning, and social networking. Apart from arising out of single discoveries or inventions, development can arise out of scientific movements or programs with defined goals for advancement of specific areas. Indeed, it is more usual to have development comprising small steps, following relatively small discoveries, over extended periods toward the goals of the intended areas. These areas are of broad societal concerns, such as agriculture, environment, and energy. Given next are some examples of development issues arising from the use of science and technology, along with other tools to improve human society, and the problems arising from them.

The Green Revolution, as it is usually called, took place around the middle of the past century, first in Mexico and then in India and other developing countries around the world. It is a revolution in agriculture stemming from R&D supported by many international and philanthropic organizations, including the World Bank, the Food and Agriculture Organization, the Rockefeller and Ford Foundations, and governments of various developed and developing countries. As we saw in Chapter 1, Norman Borlaug and others undertook extensive plant breeding, resulting in improved varieties of wheat, rice, and other major crops that were deployed in Mexico, India, the Philippines, and many other developing countries. Selection was mostly done by examining the characters of crosses between varieties (called phenotypes), helped by the use of genetic markers. These new varieties, such as semidwarf rice, which could produce higher yields without tumbling, were responsible for the dramatic success of the Green Revolution, supported by improved irrigation, extensive use of pesticides, and chemical fertilizers. Cereal production more than doubled in just two decades. Apart from Africa, in which various social infrastructure and geographical problems could not be overcome, supporters of the Green Revolution have pointed out that it has helped many developing countries avoid famine and increase their food security. Hence a strong case can be made for positive development resulting from genetic and other sciences in the Green Revolution.

However, the success of the Green Revolution did not come without serious criticisms. Increased productivity required a high input of chemical fertilizers, which caused environmental damage in terms of water quality and which contributed to global warming through the production of greenhouse gases. Chemical fertilizers are also oil-based products, raising uncertainty in supply when oil resources will eventually run out. Intensive farming requires large amounts of water, which could deprive those who do not benefit from irrigation. Another serious criticism is that deployment of only a few high-yield varieties is a serious threat to world biodiversity, since various native species of cereal and other crops are abandoned and thereby come under the threat of extinction. These criticisms have been taken into account by people such as the Indian geneticist M. S. Swaminathan, who adopt approaches that do not pit agricultural productivity against ecological and other effects and integrate desirable goals together. The new movement is called the Evergreen Revolution and is one of the early efforts to achieve sustainable development, which we will deal with in Chapter 10.

Another example of development arising from the use of science and technology is the generation and use of energy. As we saw earlier, the invention of the steam engine caused massive changes in industry, transportation, and other areas, contributing significantly to the birth of the Industrial Revolution. Increased energy use also comes from heating and cooling for houses and buildings. This caused a large increase in the requirement for energy, which up to now has come mostly from fossil fuels and other sources that require burning. Carbon dioxide and other greenhouse gases are released in these processes, leading to global warming and other unwanted environmental effects. These problems, together with future depletion of oil, gas, and other fossil fuels, have led to the search for alternative sources of energy. Renewable energy sources, including biogas, solar energy, and wind, are being used increasingly, reaching a figure of around 20% in 2015 [8] from what was insignificant just a few years ago, thanks to new technological developments for all these sources. Nuclear energy is another potential source, but perceived risks of accidental leaks of radioactive materials to the environment have put a halt to its widespread deployment in the foreseeable future. While current nuclear technology obtains energy from fission of radioactive materials, there is a new technology on the horizon based on fusion of hydrogen atoms into helium, which is free from radioactive contamination.

Concerns about climate change come from realization of detrimental effects of human activities, such as deforestation for housing, forest products or fuel, and agricultural practice that generates the greenhouse gas methane. By far the largest contributor to climate change is the burning of fossil fuels in industry, transportation, and household activities. These activities result in the release of carbon dioxide, which traps heat from the sun in the atmosphere. This and other gases released from industry, transportation, agriculture, and other activities are responsible for the greenhouse effect, or warming of the global atmosphere. Over the past century, the mean global temperature has risen by about 0.8°C [9], and if allowed to continue rising, it would have disastrous consequences, such as changing weather patterns, causing storms, floods, droughts, hot spells, melting of the polar ice caps, and a subsequent rise in sea levels. These concerns lead to a different green movement, energizing large numbers of people from all walks of life to unite in attempts to protect the environment from irreparable insults from human activities. Although there is opposition from the "business as usual" groups, the global movement is politically and socially strong and has seen development of new industries with products and processes that are kinder to nature. For example, there are now strict standards of pollution from industries and homes in many countries. Green products are popular both for homes and for businesses, such as organic foods, biogas, and biodegradable plastics. Nations have come together to pledge cooperation in limiting greenhouse gas production so as to try to control the rise in global temperature and consequent unwanted effects.

Science and its technological sparks have resulted in the development of many areas, drastically changing the way we live and work. Each development, beneficial as it is to society in general, is not free from problems. Arguably, science and technology as a main factor in changing our way of life in modern days have inadvertently caused many problems such as climate change and pollution (Fig. 2.3). However, science and technology are also essential tools for repairing the damage done and preventing future problems. The situation is similar to the Eastern proverb "A thorn inside the flesh can be taken out by another thorn." Therefore, rather than going forward with development in the narrow sense of achieving a specific purpose without due regard for the side effects, science and technology should be aimed toward not just specific development but also sustainable development.

**Figure 2.3** Benefits at a cost. Science and technology have produced many benefits for humans but have also contributed to climate change and other unwanted side effects for the earth.

## 2.7 Sparks toward Sustainability

We have seen that the sparks from the spirit of science, in the form of knowledge about nature, and ideas and innovations stemming from them, have led to development in many areas. However, we see also that such development, be it in agriculture, industry, or lifestyle, often leads to undesirable side effects. Many of these are so serious that they render the development unsustainable, that is, it cannot be sustained without mitigation or adaptation. We can divide the problems of sustainability of development into two main groups: those that are associated with the development process as it occurs and those that are associated with livelihood and lifestyles of people, which push the earth toward or past what can be called planetary boundaries. The former group is mostly specific development processes that are imperfect and can be improved or changed altogether, such as development of plant varieties so as to rely less on water and fertilizer input, which would address a main criticism of the early stages of the Green Revolution. Development of industrial processes that require less energy, use renewable energy, or produce less pollution is another example. The latter group of sustainability problems is broader in scope and concerns human activities that, due to increasing numbers and demand for better standards of living, are depleting the earth's resources and harming the environment at alarming rates. Simply put, human consumption and material production, much of it due to the increased power of science and technology, cannot increase indefinitely on a finite planet. There are planetary boundaries [10] that humanity should collectively make sure in not exceeding, as otherwise devastating consequences would result both for this and future generations. These boundaries have been identified as, to name but a few, climate change, ocean acidification (due to carbon dioxide absorption), stratospheric ozone depletion, freshwater use, land use, and chemical pollution.

These problems call for humanity's collective action, in addition to individual development efforts in selected areas. The past two decades or so have seen investigation and development of new knowledge on systemic effects of human economic activities and interactions with nature and the environment, with a view toward achieving sustainability. Sustainability science, as this new area is called, should go a long way toward helping us reach a stable situation, where humans can live prosperously in the long term on this finite planet.

# Chapter 3

# Nurturing the Spirit, Enhancing the Sparks

> *To learn to read is to light a fire; every syllable that is spelled out is a spark.*
>
> —Victor Hugo, *The Hunchback of Notre Dame,* 1831

The spirit of science needs nurturing, as otherwise it might weaken due to outside forces from society, which tend to distract young minds. This spirit, together with the inventive sparks, can be nurtured and enhanced by a system of education such as one that integrates science with technology, engineering, and mathematics (STEM). Such a system should encourage learners to be able to both learn from various sources with critical judgment and gain knowledge of the world around them. It should be integrated further with social science and humanities so that learners have a sense of perspective of where science-related areas stand within society. The system should produce people who are creative, critical, and investigative and have communicative and social skills. Various methods to bring about these desired goals include e-learning, educational gaming, and online laboratories, but these must have active guidance from competent teachers and be based also on real experience. Success depends not only on programs in classrooms and schools but also on the participation of all stakeholders, including the family, the neighborhood, and the social environment.

*Sparks from the Spirit: From Science to Innovation, Development, and Sustainability*
Yongyuth Yuthavong

ISBN 978-981-4774-57-4 (Hardcover), 978-1-315-14599-0 (eBook)
www.panstanford.com

## 3.1 Keeping the Spirit Alive

We recall that the spirit of science, embodied in curiosity, wonder, and exploration, is strong in children, but it tends to fade away or just stay latent as we grow up. For some, scientists, technologists, and innovators included, the spirit never goes away, and indeed it is crucial for the success of their careers. For others such as lawyers, businesspeople, and artists, the spirit may spring up as an inspiration for solving problems or launching new projects. However, for many others, many more than these people, the spirit is not called into action in their daily lives and may be lost or forgotten. Yet it is surely good to keep the spirit alive, not for the sake of science, but for our own guidance in various undertakings. Recall the "six honest serving men" who taught Rudyard Kipling all he knew, in Chapter 1: why, when, where, how, what, and who. They also teach us all we know. The collective spirit of these six honest men is akin to the spirit of science and should be invoked so that we can do well in our daily lives as well as our jobs (Fig. 3.1).

**Figure 3.1** The six honest serving men according to Rudyard Kipling (*The Elephant's Child*, 1902).

Everything around us in daily life keeps these six serving men busy. We have to go about our daily routines, eat and play with our families and friends, make short- and long-term plans for ourselves and our families, build our careers, and think about good ways to save and spend money. True, we can do all this without calling these six men into action, but it is surely better to have them work for our sake. Take just the question of saving money. Why do we need to save? When and where should we do it? How can we save money appropriately and not become too stingy? What aspects of saving should we concentrate on? And in the family, who should do more saving, the children, the spouse, or you? These are the questions seemingly far from those posed by scientists in their work but, indeed, are the questions that make us undergo the same type of thinking and undertaking as scientists. The difference is that the questions asked by scientists concern phenomena in nature that they would like to understand in all important aspects. Let us change the topic from personal saving to the scientific topic of climate change and use the service of the six honest men. Is climate change occurring? If so, why? When and where has it occurred? How can we predict the course of events, and how do we mitigate its effect or make appropriate adaptations of our lifestyle? What are the effects of climate change in terms of influence on the planet, people, and living things in general? Who are the people affected, and who are those who can help to reduce the effects of climate change? The same types of questions can be asked for various issues in science. The difference from normal questions in our daily lives is that they usually have a higher level of complexities, and attempts to answer them need to be made in a systematic manner. This involves experiments and observation by various groups of scientists, who will communicate their findings after the results they obtain are subject to critical reviews so as to advance the state of the science further together as a global community of scientists.

What about the sparks from the spirit of science? In the case of climate change investigations, the sparks may range from ideas about devices that can capture greenhouse gases and store them away from the atmosphere, to the development of an international accord to limit global temperature rise, to the replacement of fossil fuels by renewable energy that would result in sustainable energy use. As for our little example about personal saving, the sparks could well be engagement in new types of saving that bring back more returns with reduced risks. There are many open possibilities once we are engaged in rational investigation with the spirit of the six honest men.

How do we nurture the spirit and encourage the sparks throughout our lives and not just see them fade beyond childhood years? This depends much on the environment in which children grow up and adults work and live. Interactions within the family, within the community, and with society at large; the constructive role of the mass media; encouragement and support of the civic and political institutions, all these play important roles in shaping the views and actions of the members of society. Most important of all is the role of education, to which we now turn our interest.

## 3.2 The Role of Education in Nurturing the Spirit and Enhancing the Sparks

The aim of education is to gain knowledge, experience, skill, and sound attitudes that contribute to building good character. All of these contribute toward success in our jobs, fulfillment of our duties and responsibilities, happiness in our families, and a healthy relationship with others and the world around us in general. Knowledge can be defined as awareness, understanding, and familiarity with facts, information, and skills acquired through perceiving, discovering, and learning. We obtain knowledge from many sources and through many channels. At schools and other places of learning, we get knowledge from our teachers, who teach us in classrooms with the help of books and other educational aids. We also go through informal learning through personal interactions with our parents, other members of our families, and other people we know. Often, we are first exposed to information from various sources before we obtain what can be called knowledge. In the past few years, social networking has become an important source of information, albeit undigested and often untrue and misleading. Another source of information is websites and mass media, which also need verification and critical judgment. The true mark of successful education is the ability to find and sift through information from various sources, analyze and compare their contents with our prior knowledge, consult our teachers and other people, and then come up with our own synthesis of new knowledge. Education can bring us to another level of intellectual capability that we might call wisdom. This is the quality resulting from having a large body of knowledge, experience, and critical awareness leading to good judgment, prudence, and circumspection. Such quality requires both the ability of the brain and the spirit of the heart, of which the spirit of science is a part. Without good education, we will only be exposed to information without gaining knowledge, let alone wisdom, as noted by T. S. Eliot [11]:

*Where is the wisdom we have lost in knowledge?*

*Where is the knowledge we have lost in information?*

Good education in science comprises an understanding of the basic principles, an awareness of scientific information, the ability to put the information in the context of the principles, and the ability to appreciate the implications of the information, including potential problems such as accuracy and significance to the field. Importantly, this basic background should stir up a number of questions from our curiosity and desire to learn more about the subject (Fig. 3.2). For example, in the science of energy we need to know the basic principles concerning energy of various forms and their transformation, including examples of such transformation in motors, dynamos, and heating and cooling instruments and the mechanisms by which all this takes place. These ground rules will allow us to understand the working of various machines that transform energy, or utilize the machines for some purpose. We should also know the principle that efficiency of transformation can never reach 100% and some energy is always lost as heat. Implications of this knowledge include our ability to judge the quality of various machines from their energy efficiency, and our appreciation of the broad picture that the more energy we use, as is the case with the present state of the world, the more heat will be produced, contributing to global warming as a major side effect. We should then be able to ask a few questions to ourselves and our teachers and friends on the present status of the aspects of energy in which we are interested and be able to search for information by ourselves from books, the Internet, and various sources.

**Figure 3.2** Good science education should not only give background information but also bring up questions for further study.

Skills and practical experience are important parts of education, especially for science and other technical subjects. Good schools and institutes of higher learning therefore need to be equipped with facilities to allow students to gain practical knowledge in various scientific areas. It is important that students have hands-on experience, not just learn from books or witness demonstrations, which by themselves are useful but not enough. Realization and appreciation of scientific concepts and acquisition of technical skills can only come from real practice in observation, exploration, and use of tools of science and technology. Vocational schools, technical colleges, and universities are the places to acquire the skills and experience, but they need not, and should not, be the only places. There are now many programs of cooperation between these institutes and production and service firms for work-integrated learning so that technical and engineering students can acquire work experience and relate it to their classroom lessons. Applying skills and experience to the workplace is not a routine process; in reality there will be problems with the machines and processes and the need to adapt them to situations such as changes in raw material supplies, machine malfunction and services, availability of spare parts, and replacement procedures. Good educational institutes therefore also prepare the students for solving problems and further for adapting to or even innovating processes, hardware, and software. Practical experience and skills are needed not only for engineering tasks but also for all areas of experimental science and technology. A chemist, for example, needs to be able to handle various chemicals, be familiar with chemical processes, and be capable of working on synthesis and analysis at the bench. A marine biologist needs to be familiar with various animals and plants in their marine habitats and have necessary skills for survey and analysis of marine species and their environment. A computer scientist should have basic familiarity with and skills in working with hardware and software of his or her specialty, including design and troubleshooting, and not just know the principles of semiconductors and computing.

The basic skills and experience of scientists and technologists are the stepping stones to go beyond the boundaries of current knowledge, that is, to explore unknown areas and utilize new aspects through research and development (R&D). Good education should not only instill in the learners current knowledge and skills in the field but also enable them to ask new questions and to try to solve them through observation and experimentation. It should also encourage them to have the sparks from the spirit of science and technology to go on to innovations of practical benefits to society. We return to Kipling's six honest serving men mentioned in the last section, but this time to use them not only to do the already known tasks but also to go beyond the current status of knowledge so as to discover new principles, products, and processes and to utilize them further through innovations. An example of new discoveries that extends current knowledge might be that concerning genes. We already know that genes, as strings of the four-letter alphabet of deoxyribonucleic acid (DNA), code for various characters of living beings. However, we only know the function of a small fraction of the language of the DNA alphabet in the cell, namely what the genes code for, leaving the rest that comprises the majority of the DNA to be discovered. New frontiers include knowledge of which parts of the genome are utilized, when, and for what purposes. Such knowledge can be used to light the sparks of innovation on diagnosis, prevention, and therapy of diseases, for example. Another example of new innovation might be development of new machines or materials that can improve the efficiency of conversion of solar energy into electricity. Figure 3.3 is a diagram showing the relationship between education and R&D, which is reciprocal in nature. Good education leads to R&D through prior knowledge foundation, and good R&D leads to extension of knowledge as materials for education. The intellectual and material resources needed for these activities are paid back in terms of societal benefits.

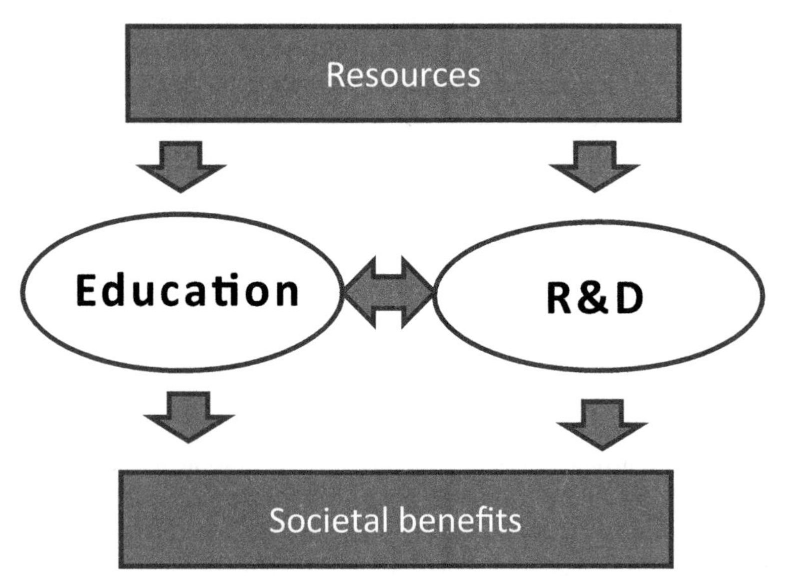


**Figure 3.3** Reciprocal relationship between education and research and development (R&D). R&D requires intellectual and material resources as input and results in a body of knowledge and applications as output. Education adopts this body of knowledge and applications as raw materials, in addition to providing an already established body of knowledge and applications for R&D purposes, with societal benefits as the outcome of these two interacting activities.

The skills needed in good scientific education and for going on to successful tasks in later life, such as R&D, are parts of a broader set of skills generally referred to as 21st-century skills. The new century calls for basic skills, some of which have been recognized as important even in past times, but many of which are new and called for in response to the emergence of modern society with its new sets of problems and capabilities. These skills are needed for happiness and success in the workplace, in the home, and in living in general. They are also essential for nurturing the spirit of science and enhancing the sparks from it. "Partnership for 21st Century Learning" [12] lists three types of skills, namely learning skills, literacy skills, and life skills. Learning skills include critical thinking, creative thinking, collaborating, and communicating. Literacy skills include information literacy, media literacy, and technology literacy. Life skills include flexibility, initiative, social skills, productivity, and leadership.

Good education leads to the formation of good character, which includes having sound attitudes toward society and the environment, an open mind, tolerance of dissent, willingness to be engaged in inquiry and investigation, and passion and willingness to devote efforts to obtain relevant answers. Different societies may have somewhat different concepts of "sound" attitudes, but in general they are about having, on the one hand, balanced considerations for ourselves, our family and friends, our countries, and the society at large and, on the other hand, balanced considerations for humans and the environment. An open mind and tolerance of dissent are consequences of a good education, which opens the learners to different interpretations about events or issues and leads to the realization of the importance of reasoned discussions and evidence-based conclusions. Passion and willingness to make efforts to become familiar with, and eventually to master, the subjects of learning are parts of a good character. This will benefit not only scholarship but also other aspects of the learner's character, such as passion for jobs and duties and for justice. Altogether such passion should help the learner become a better person and to try to help building a better world. Indeed, many people regard formation of character as the major objective of education. This is a good judgment since, while knowledge and skills may be forgotten as time goes by, character stays with us all our lives. The spirit of science entrenched in this character can then be nurtured, and the sparks from the spirit enhanced as new ideas, innovations, and other novel undertakings.

## 3.3 Science, Technology, Engineering, and Mathematics as Content

What should be the content of successful learning to cultivate the spirit of science? The basic principles and information from important disciplines of science and technology are essential, but for students to be inspired and motivated there should be more than just factual content or formulae for derivation of equations or for solving them. It is more important to stimulate the interest of students through examples that come from real-life situations, from common curiosities about nature, or from the ability of mathematics to explain phenomena like the movement of stars. Conventionally, the content of learning at school levels is distributed among various subjects, like chemistry in science and geometry in mathematics. By tradition, little school learning dealt with technology and engineering, which were mostly left until the university level. In the past few years, however, countries around the world have established programs of learning in science and related subjects of technology, engineering, and mathematics from the school level onward. The movement to concentrate on education in science, technology, engineering, and mathematics, or STEM education as it is called [13], was widely ascribed to the initiative of the U.S. National Science Foundation in the 1990s and in the policy commitment of President Barrack Obama in 2009. However, the roots of this movement go back a long way, including major events in the second half of the past century. The space race between the United States and the USSR highlighted the importance of education in science and related subjects, both for these two countries and for others around the world. The world paid attention to the triumph of the USSR in sending the first satellite, *Sputnik*, into orbit in 1957. It paid attention when the United States managed to put men on the moon 12 years later. These events, as well as important discoveries and inventions based on science and technology at an increasing pace, made it evident that science and technology are the main factors for the success of economies, enhancing their competitiveness and resilience. Newly industrializing countries like those in East Asia were quick to see the importance of good systems of education in science and technology for the success of their economies. STEM education has therefore been adopted in many countries, both developed and developing, as a major tool to upgrade the quality of their human resources. International comparison shows that there is a general correlation between students' achievements in science and mathematics and per capita income, underscoring the importance of good science and mathematics education in national development.

One main feature of STEM education is that it can introduce the students to integrative learning. Students do not just learn subjects in STEM as separate disciplines but also learn about their connections and synergy and can develop a higher level of thinking skills. For example, in learning life sciences, students learn about not only descriptive life forms but also their underlying mechanisms and functions in terms of physics and chemistry, the technologies for analysis and synthesis of biological molecules, engineering aspects of living beings and instruments to investigate them, and mathematics behind the physical techniques. In learning about the physics of light, students get to know both the physical principles of light, mathematics of waves, and engineering and the mathematical aspects of equipment to detect and transform light. Often, integration involves project-based learning, that is, learning not just through the disciplines but also through real projects that require problem solving, such as survival on the moon or exploration of deep oceans. STEM education can give broader perspectives to students in facing real problems in everyday life and prepare them better for their future. In many cases, students are encouraged to work together on STEM topics, enhancing their interactions and learning outcomes. However, for integrated STEM education to be successful, the students and the teachers need to be well prepared in separate disciplines and need to develop capabilities in referring back to the principles of different disciplines. There is also a danger of weak students in the group being left behind. This requires good preparation for teaching, often involving teams of skilled collaborating teachers. This makes it difficult for small schools, especially those in developing countries, to carry out effective STEM education. STEM education has also been adopted by many countries for their universities and technical colleges, since it is a good preparation for jobs in the modern sector, including information technology, medicine, biotechnology, and materials science. In these areas, while the newly industrializing countries are doing well, there is a danger that the developing countries in general will lag behind because of problems in adapting from a conventional to a more modern STEM-based education.

We tend to think of education as a formal process of learning in schools or institutes of higher learning, and the teachers as the main conduit for transfer of knowledge. In fact, we learn from many sources other than schools and institutes, such as from our families and friends, our communities, and the world at large through the news, social media, the Internet, and many other informal sources. Young children gain so much fun from going to science museums and other exploration venues, where science is on display and science games offered for their participation. Online exploration sites can also give exciting virtual learning experiences and are valuable for young people without access to science centers, such as in many developing countries. They also enjoy events such as science camps in museums or outdoor venues where they can explore science together with friends. International collaboration can give young people broadened views of science, which is a pursuit of knowledge without borders. Box 3.1 shows examples of fun activities, such as the Asian Science Camp, with participation of young people from all over Asia.

**Box 3.1** Building the Spirit through Fun Activities

Science is not all about hard learning with no room for fun.

On the contrary, the spirit of science often starts with fun activities, with learning as a by-product. We enjoy playing with toys, from tops and slides to Rubik's Cube. Only later do we realize, or are told by our teachers, that the principles of physics apply to our toys. For the curious and exploring ones, these are the triggers for growth of the spirit of science. Activities such as trips to science museums, exploratoriums, and planetariums and field trips to explore nature are all fun activities that we remember with fondness as we grow up. Many are so impressed as to make a career choice in science, or at least give science an important part in their lives and their decisions as they grow up. Such activities, guided suitably by teachers, trained staff, or volunteers, allow children to experience the wonders of science and play or experiment with concepts and devices. This active learning is accompanied by simplified explanations pointing out the main scientific principles involved, without too many details that could deter the learners and dampen their enthusiasm.

These activities can be held at various levels, as classroom outings, as school night-outs at museums, or as science camps and festivals with participation from many schools or even schools and individuals from different countries. The Asian Science Camp, for instance, is an international event jointly organized by many Asian countries on an annual basis. Some organizations run these activities as part of programs to recruit young people into science careers or to ignite and retain their interest in science irrespective of their later career choice. As an example, the National Science and Technology Development Agency of Thailand, runs the Sirindhorn Science Home, named after a princess who is an active supporter of science, with year-round fun activities on science and related active learning, some of which are done together with research scientists in the agency. Recently, an Asian Science Camp was organized there with participation of some 270 young people from 30 countries, who showed promise in science at the high school and university levels, with inspiring activities, including discussions with Nobel laureates. At a more advanced level, once every year, over more than six decades, dozens of past Nobel laureates have convened at the Lindau Nobel Laureate Meeting in Germany to meet and exchange ideas with leading young scientists, postdoctoral researchers, PhD students, and undergraduates. These activities will make long-lasting impressions on the next generation of scientists and technologists to ignite their spirits and the sparks from them.

These sources are utilized by STEM education to increase the effectiveness of learning. Furthermore, some countries, such as the United Kingdom and the United States, have initiated STEM ambassador programs, recruiting college students, working engineers, scientists, and other professionals, as well as businesspeople, to encourage young people to enjoy STEM subjects and to inspire them to build careers based on STEM experience. Through this approach, young people can acquire the spirit of science, not just from encouragement of people who are already scientists and technologists, but also from people from all walks of life.

## 3.4 Society and Environment as a Context

However important a good education in science and related areas such as those in STEM may be, it should not constitute the sole content of what is needed for young people to be equipped to have successful careers, to lead happy lives, and to play constructive roles in society. Most people do not intend to have careers in science and technology. For them, as well as for those who would like to be scientists and technologists, it is important to know the role of science and technology in societal and environmental contexts. The society and the environment begin with our families and extend outward to our neighborhoods, villages, towns and cities, countries, and the world at large. On the society side, we are concerned with issues such as the ways of life, including gender issues, urbanization, health, entertainment and culture, the arts, tourism, democratization, laws, and religious beliefs. On the environment side, we need to pay attention to issues concerning urbanization, forest conservation and biodiversity, water access and quality, oceans and seas, pollution, natural and man-made disasters, and the looming climate change. In addition to these social and environmental issues, we also need to be aware of economic issues that play crucial roles in our society, including jobs, income and poverty, trade and industry, agriculture, transport, and information and communication services. Finally, we also need to be aware of issues concerning security, political, and military conflicts and accords within and between countries. These issues extend beyond the immediate concerns of science and technology but are important for all students, no matter what careers or styles of living they intend to adopt.

Many educational institutes offer these various issues as subjects for degrees in liberal arts, providing a broad choice for students to take. Liberal arts graduates are broadly educated, in accordance with the classical tradition that an educated person is a free (liberated) person. Traditionally, liberal arts graduates tend to be proficient in areas such as literature, languages, art, music, and history. However, liberal arts also include science and mathematics, albeit in broader scopes and less detail than STEM. Many people therefore take the view that there is a dichotomy between the two and that people with a serious intention to pursue science and technology should take STEM rather than liberal arts education. In light of our conclusion that the societal and environmental context of education is important for everyone, including those who intend to have scientific and technological careers, the liberal arts aspects should be included in a science-based curriculum, and for those who intend to have a liberal arts education, STEM aspects should also be included. Educational programs should be tailored according to the need and intention of the students so that a good balance is made to ensure both proficiency in the professions and a broad awareness of the context of their education.

## 3.5 Nurturing the Spirit: Absorption and Building of Knowledge

"We are what we eat." The statement in Fig. 3.4 is true, but only partially so. The food we eat is first digested to its basic components, from proteins to amino acids, from starch to sugars, and from fats to fatty acids and glycerol. These basic components are used within the cells or transported to where they are needed for replenishment. They are then reassembled into larger components that make up the structures of our cells and tissues and used for giving energy as we need for our various activities. Although the basic components of our bodies are the same as those in the food we eat, they are entirely different in form and function. We have transformed our food and constructed our own bodies—muscles, bones, brain, and all—from the basic components in our food.

Similar things can be said about learning. Although simplistically we may think that in learning we take in what is taught to us by our teachers or from textbooks, manuals, and other sources, through memory and practice, the reality is far more complex. Educational psychologists have shown that learning is an active process that involves not only absorption of information but also reasoning, reflection, integration with previous knowledge, and experience and synthesis of our own new knowledge. The environment, our motivation, and state of mind during learning, inspiration from the teachers, and the sources of learning, all play important parts in learning. There are many schools of thought on learning (Box 3.2), some emphasizing a change in behavior and our understanding of the world around us, others concluding that we construct our own knowledge of the world from the information we take in, which is put into the context of what we already know and believe, and still others pointing out that learning is a transformative process that redefines our view of the world. A common aspect of these various lines of explanation of learning is that it is an active process of absorption of new information, from which we discover and build our own system of knowledge and skills, influenced by our innate interest and preconceptions. Just as we digest the food we eat and use its basic components to build our own bodies' energy sources, so must we in effective learning digest and absorb the new information, integrate it with our already existing store of knowledge, and synthesize new knowledge or develop our worldview.

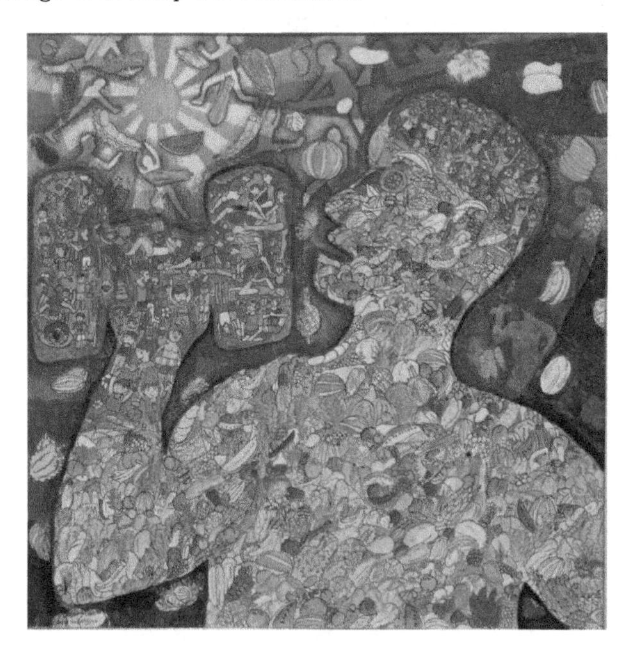

**Figure 3.4** "We are what we eat." The statement is only partly true, since although the food we eat becomes our building blocks, it has to be broken down into basic components and reassembled. The process of digestion and reassembling of the basic ingredients is akin to education. In good education, information and practices are transformed into knowledge and skills, a spirit for learning and exploration, and a healthy mind, just as the transformation process in nurturing our body. *Healthy Body*, a painting by Kasha Kamdam, World Second Prize in Drawings & Paintings Category, 9–13-year-old age class of PMAC 2016 World Art Contest, activity under the Prince Mahidol Award Conference 2016.

**Box 3.2** Learning as an Active Process

It used to be taken for granted that we learn what we are taught. Therefore, the thinking is, "If you want the student to learn more, you have to teach more." In many countries, the traditional style of learning—learning by rote—is still widely practiced. Some students, who have a good memory, may be able to remember what they are taught in class or from books, but most have difficulty. For all students, although they may be able to remember some content, they can only partially understand it, if at all. This is especially true of science, which requires reasoning in order to be able to understand. Worse still, the students are usually turned off from the subject, branding it as boring and too time-demanding.

Thankfully, we now understand more about the nature of learning. Far from being a process where the teacher is actively teaching and the learner is passively taking in the content, it is the learner who is the more active partner in the process. It is now generally agreed that learning is an active process, whereby the learner must absorb, process, and retain the information provided from outside, be it from a teacher, a book, or the Internet. Paradoxically, the learner learns more when the teacher teaches less, although guidance from the teacher is of course needed. An important step in learning is understanding the information and putting it in context in the learner's mind, the "Aha!" moment in learning science or other challenging subjects. This requires attempts by the learner in reasoning and practice (in skill learning), backed by emotional support, that is, the will to learn, followed by an appreciation of success in learning as a reward. Successful learning leads to a change in the behavior and world view of the learner. In analogy with the digestion of food and building our muscles from it, the learner digests the information from outside and uses the materials to construct his or her own muscles. Jean Piaget, the founder of the theory of learning, called constructivism, argued that humans generate knowledge and meaning from interaction between their experiences and their ideas. Seymour Papert argued further that learning happens most effectively when the learner constructs a meaningful, tangible product in the process, which he called constructionist learning. Constructionism is a powerful theory of learning, especially in the context of learning science and technology, with importance given to hands-on experience, problem-based learning, and synthesis of knowledge from observation and experiments. We are reminded of Confucius (551–479 BCE), who said, "I hear and I forget, I see and I remember, I do and I understand."

In learning science, many students already have innate interest and curiosity in knowing more about the wonders of nature. This gives a good start for effective learning, which can be achieved through STEM or other constructive means of learning, as already discussed. Effective learning and enthusiasm for learning more are mutually reinforcing. A learner may start from early childhood with a spirit of curiosity, wonder, and exploration innate in all of us. A good education should allow the child to build knowledge about nature and skills for its exploration and, just as importantly, build the enthusiasm and motivation for learning more. As an example, a child may be curious about the way ants follow one another in a trail. A teacher or a family member may guide the child about the nature of ants in working together as a group, each leaving a scent for others to follow so that they can bring food back to their home. As comparison, in the fairy tale "Hansel and Gretel," the two children left a trail of pebbles as they were taken away so that they could find their way back home. An experiment can be done together, rubbing out the scent in the trail with a finger or an eraser, to show that the ants are now at a loss to decide where to go. Many more examples like this, linking the world around us or man-made objects with explanations, experiments, and further questions appropriate for the level of child development, should fuel the motivation and enthusiasm for learning more. As the child grows up, the practice of asking questions and trying to obtain answers by oneself should become a habit. Granted, not every child will be so motivated, since everyone has different innate interests, but at least those who are so inclined will be attracted to the spirit of science.

There are many ways of learning and teaching science, especially today, when new ways of communication are available through information and related technologies. Classes need not have rigid structures as in the past, where they were treated like sacred places in which formal teaching took place. Instead, the flipped classroom can be established, where formal learning can take place in the home through lectures and other instructions by video, audio, the Internet, or other formats, and the classroom is turned into an informal place of group discussions, exercises, question-and-answer sessions, and project design. A number of good online courses are freely available, offered as massive open online courses (MOOCs) by many universities and learning places. The Khan Academy is an example of an institute offering popular online science courses, geared toward good, easy-to-follow content and guided by friendly teachers. The flipped classroom is one example of new models of learning and teaching, both in science and in other subjects, which achieves greater efficiency through flexibility built into the system. Other flexible systems such as competency-based or interest-driven learning pay attention to the individual, taking into account his or her intrinsic speed or interest in learning and allowing the student to discover his or her own preferences.

Science is a hands-on subject based on experiments and observation. To nurture the spirit of science, learning based on subject content alone is not enough and can even be discouraging. Students can be motivated by problem-based or project-based learning, in which problems and activities from real-life situations are the center of active discussion and investigation in the field and school laboratories, both alone and in groups. Some of the topics for such problem- or project-based learning include plant and animal development with experiments on DNA manipulation, solar energy utilization with experiments and observations on solar energy conversion to heat and electricity, and music making with experiments on vibrations and waves. Practical classes can thereby be linked to subject content taught in class. At higher levels, students can be introduced to work-integrated learning, in which the learning is done through practice in the workplace, such as in manufacturing factories or in firms linked to the learning institutions.

## 3.6 Enhancing the Sparks: Critical Success Factors

We turn now from learning, absorbing what other people have found out, to the subject of innovation, thinking, and making of new things ourselves. The ability to innovate results from a combination of several skills, some naturally endowed and others having to be acquired. It requires creative skills in spotting the need for innovation, which may be for solving problems or finding new and better solutions to, say, a household gadget, an industrial process, or a social network system. It requires technical skills in design and experimentation, going from the concept to a prototype and finally to the commercial stage. It requires managerial skills in planning, organizing, and running a team of capable people to go through with the project. In many cases, it requires financial skill and skill in risk taking, since many innovations require the raising of capital. These various skills are rarely available in one person, although many are gifted with one or more abilities. Some people are gifted with the ability to introduce new concepts or see potentials of new methods or tools to solve existing problems, while others are quick to see possible technical improvements to a machine, and still others have natural managerial skills for doing new things. Finding the right team or partner to complement one person's skill repertoire is an important step for successful innovators. The complementary abilities of Steve Jobs and Steve Wozniak in developing the Apple computer is a classic example: both were inventors, with Wozniak specializing as an electronics engineer and computer programmer and Jobs as a technology entrepreneur. In science, the successful partnership between Francis Crick, a physicist, and James Watson, a biologist, resulted in their discovery of the double-helix structure of DNA.

What are the most important aspects for enabling innovation? The various innovative skills can be used productively to produce innovation in science and technology, or indeed in other spheres of activity, but the skills cannot be deployed without an important guide—imagination. Imagination is an important quality for all creative endeavors. It allows us to dream and conceive of various new possibilities. Many ideas cannot go further, because they are out of line with scientific principles or too advanced for the present stage of knowledge, but some creative ideas can prove to be useful and can be realized as innovations. Some ideas may be beyond the limit of knowledge of the time and only come to reality later with increased knowledge. An example is the idea of Nikola Tesla, a contemporary of Einstein and the inventor of the radio transmitter and many other electrical devices, who foresaw the days when people would be able to communicate with one another anywhere in the world with ease, a vision that has become a reality only recently. Other ideas for innovation could come to fruition with support from current knowledge. Therefore, even though imagination is not limited, the next crucial step is to explore possibilities for its realization with the present state of knowledge and other limiting factors. Tesla said that when he got an idea, he would build it up in his imagination and make improvements and operate the device in his mind. Most of us do not have the gift of innovation like Tesla did, but we can start similarly by conceiving ideas and proceeding further to potential reality.

Vision is another quality closely related to imagination, important especially for conceptualizing the end results of the innovation. While imagination is open ended and allows room for free thinking, vision is a foreseen realistic situation if and when the idea bears fruit. It is an important step in creativity. Typically, the innovator might start with free imagination and then narrow it down to a more realistic vision. Both imagination and vision could come from inspirations from various sources, including teachers, encountered success models, and long-lasting impressions and aspirations.

Imagination and vision form only the beginning of a long process of innovation. Another quality needed for successful innovation is the ability to explore various aspects of the imagined concept or device. We may not be able to see the advancement from idea to reality all in detail in our mind, but we can go to various sources of knowledge, consult with people in the field, and explore the value of our potential innovation with potential users. The people we consult with should include not only technical experts but also potential users, people familiar with production and marketing, and other stakeholders. Experiences of others as well as accounts in the literature concerning similar undertakings are valuable, as most imagined innovations turn out not to be completely novel, and the history of success or failure of similar projects can be learned.

Once we have passed the stages of imagination, vision building, and exploration and gained confidence with our prospective innovation, we can call the various innovative skills described earlier into play, including risk assessment, financial management, and technical management. An important step is the production of the prototype, or production at a small scale, the success of which should encourage us to plan and execute the project with more confidence and gain more support from outside sources. The main thrust for all these steps is our passion to go ahead until the imagination or vision has been fulfilled, bearing in mind that failure can occur at any stage in the process. Even with failures, which are common in innovation projects, we should not be easily intimidated. Stamina and persistence, even in the face of failure, are critical success factors, since most projects have problems, many of them serious ones, at one time or another. With each failure, we should not just go on relentlessly but should pause and investigate its causes and ways to prevent it or mitigate its effect. However, if we meet with repeated failures and cannot find ways to correct them or cannot find alternatives, at some stage we may need to face the possibility of a failed innovation. In these cases, it is better to find out sooner rather than later.

We have concentrated on enhancing the sparks of innovation from the spirit of science. This occurs mainly at the individual or the organizational level. As we saw earlier, the spirit of science can help ignite sparks at higher levels, namely development and preferably sustainable development. In these cases the sparks can be ignited or enhanced, not just by one or a few individuals, but by movements started or powered by large sectors of society. The movements to reduce and mitigate the effects of climate change, for example, do not come just from a few individuals but are the results of long processes of studies, debates, and policy decisions at various levels. However, in these large movements, the crucial roles of a few individuals can be identified. The environmental conservation movement in the 1960s, for example, can be traced back to the influential book *Silent Spring* by Rachel Carson. The Millennium Development Goals, a set of ambitious goals described as a road map for world development up to 2015, had Jeffery Sachs as the director of the project. These were followed by the Sustainable Development Goals to be accomplished in 2030, adopted by UN member states and supported by a large number of international organizations. These social, environmental, and technological development movements are the results of a consensus of large numbers of people who agree on the problems and needs for their solutions but require a number of dedicated people and organizations to work on them in a prolonged and sustained manner in order to bring success.

# Chapter 4

# Cultivating the Brain

*Discovery consists of seeing what everybody has seen,*
*and thinking what nobody has thought.*
—Albert Szent-Gyorgyi, *Irving Good, The Scientist Speculates* (1962)

The ability to think is innate, but in humans this can be cultivated or destroyed by upbringing. The brain can be stimulated and maintained throughout life. Science helps cultivate the brain, as it encourages reasoning, imagination, and creativity. The ability to think creatively is closely associated with the ability to ask questions and to proceed to find answers to the questions. Discipline is needed so that established rules and principles are respected. At the same time, the ability to think out of the box and to rebel against prevailing concepts is also needed, opening the way to new ones that may prove to be superior. Apart from cultivating the brain with the ability to think, we should also cultivate our psychomotor skills, linking thinking with manipulative capability. Importantly, we also need to cultivate passion as the motivator for action. Cultivating the brain goes beyond cultivating scientific ability and encompasses various areas including verbal, artistic, athletic, and contemplative abilities. These multiple facets of intelligence can be cultivated and combined appropriately, as is needed for the spirit of science in particular and to hone 21st-century skills in general.

*Sparks from the Spirit: From Science to Innovation, Development, and Sustainability*
Yongyuth Yuthavong

ISBN 978-981-4774-57-4 (Hardcover), 978-1-315-14599-0 (eBook)
www.panstanford.com

## 4.1 Human Thinking Can Be Cultivated

Animals can think, more or less. An eagle can think of ways to prey on a squirrel it sees, and the squirrel can think of how to run away or hide from the eagle. Highly intelligent animals like the chimpanzee can even think of how to make and use tools. As observed first by Jane Goodall, chimpanzees strip leaves off small twigs and use them to fish for termites from the ground. Even lower animals like the box jellyfish use a hunting strategy to obtain their prey. True, some expressions of animal behavior may not come from conscious thinking but more from instincts, such as mating and raising the young. Konrad Lorenz found that hatched goslings would imprint on the first object they saw, even him, and followed him as they would follow their mother. Aside from innate behavior from instinct or imprinting, many other aspects of the behavior of animals such as dogs, elephants, and whales likely come from their ability to think.

Although many animals are classified as intelligent and being able to think well, they are far from humans in thinking capability. Animals can be trained to do complex tasks, but there is no evidence that they can think in complex manners or that they can cultivate their own brain capability like humans. Complex thinking includes imagining, thinking in an abstract manner, thinking many steps ahead, thinking of arts and beauty, planning, solving problems, and thinking in a scientific way. The fact that chimpanzees can use twigs to tease out termites might mean that they can plan and think ahead to some extent but only at an elementary level compared with what humans do. Ask a child to tease out termites, and you will see many solutions, including various designs of rods and use of syrup and other substances to enhance the process. Furthermore, an important distinction of humans from animals is the former's ability to cultivate their thinking in advanced and complex ways, not just through learning in general, but also through learning to think in particular. As we saw in the last chapter, effective human learning occurs when the learner not only gets to know the content of the subject but also gets to understand, to be able to reflect upon it, and to be able to construct mental models of the subject and relate them to other things previously learned. The ability to cultivate the brain goes beyond the simple ability to think and is a special gift possibly reserved for humans.

We see evidence of cultivated thinking in humans right from the dawn of human evolution. Prehistoric cave drawings in many parts of the world show that early humans already had complex thinking, in terms of capturing of nature, artistic representation, and possibly even further to planning of hunting and gathering of food and amenities. The human ability to think increased continuously with the evolution of the brain, which became both larger and more complex in structure and function, especially in the cerebral cortex, which contains what is called gray matter. There are approximately 85 billion neurons (nerve cells), each of which is connected to tens of thousands of other neurons. They work together in concert through communication channels and also through interaction with signals from hormones and other sources. The ability to think is furthermore enhanced by practice, especially in interacting with other people and the environment. The brain is a plastic entity (Box 4.1), as it can be activated and its ability enhanced by learning, social interaction, and living in general, especially when it is challenged by new situations or problems it wants to solve. There is also physical evidence that the brain undergoes physical changes, both at the neuron level and at the whole-brain level, in learning, memory, and recovery from brain damage. The brain can be cultivated throughout life, but the most crucial period is early childhood, as evidence shows significant results of development in the intelligence quotient (IQ), social and emotional skills, and return on investment, especially for disadvantaged children, who can benefit greatly from early intervention [14].

**Box 4.1** The Malleable Brain

The brain is a plastic entity in the sense that it is malleable, that is, it can change according to an external stimulus and self-motivated learning and practice [19]. The changes include acquisition of new skills, knowledge, and abilities and the loss of previous ones when they are no longer used. The changes can come from conscious learning or from acquisition from new behavior. Formerly, it was believed that such changes mainly occur during childhood and stop or occur only minimally as we are fully grown. "You can't teach an old dog new tricks," as the saying goes, meaning that old people are not able or willing to learn new things, or even change their old habits. This may be partly true in the sense that people are used to their old ways and unwilling to change. Many people, especially elderly ones, who are used to reading the newspaper or watching TV are unlikely to change their habits to reading and watching things online. However, unwillingness to learn or do new things is not the same as inability. Elderly people who are used to old habits or reluctant to learn new things can indeed learn new things and adopt new habits. Their brains are still plastic, although not so much as when they were young. Once they have the motivation, inducement from employers, or a desire to be engaged in social networks perhaps, they can learn and develop new capabilities to a considerable extent no matter how old they are. Even people who lost their previous skills, knowledge, and memory due to illness or accidents can revive them to a certain extent, depending on the seriousness and type of disability.

The plasticity of the brain is revealed not only in behavioral changes but also physically in anatomical changes in the brain. New neural connections and circuitries can be shown to occur in learning, in children, grown-ups, or rehabilitating persons. Activities of the mind, such as mindfulness meditation, have also been shown to lead to changes in the brain. Studies on such changes in the brain, together with more knowledge in the psychology and physiology of how we learn, have led to the development of brain-based learning as an important approach to education, comprising teaching methods, lesson designs, and learning programs. The spirit of science and its sparks can gain tremendously from such an approach to learning, which can be applied to people of all ages, not just children.

## 4.2 Science Helps Cultivate Thinking

We have seen earlier that imagination helps nurture the spirit of science. Vice versa, the spirit of science helps build imagination (Fig. 4.1). Imagination is the part of thinking that occurs when we free our brain of previous restrictions and venture into the unknown, as science encourages us to do. Furthermore, science involves a systematic way of thinking, starting with asking questions and trying to find answers by offering a tentative and reasonable explanation, followed by observation and experiments that can decide whether our explanation is correct or not. Even when we cannot yet offer tentative answers, various lines of observation and experiments can sometimes lead us to answers. This scientific way of thinking and working is not the monopoly of scientists but a generally effective way of thinking and problem solving of people in general. Science only adds logic and a rigorous way of thinking and action so that good solutions or conclusions can be found.

**Figure 4.1** Science encourages us to build imagination.

In cultivating scientific thinking, which can also be applied generally to domains other than science, we must have two contrasting spirits, the spirit of discipline and the spirit of rebellion (Fig. 4.2). The spirit of discipline consists of a willingness to accept the known principles and previous findings as the basis for asking questions and formulating tentative answers. For example, in asking whether we can make more effective solar cells than before, we will need to accept previous theories and practice in making solar cells and go on from there to try to find new substances to capture solar energy using the known principles of solar science. The spirit of rebellion is the ability to break away from accepted principles and find better alternatives in theory or practice. In the example of solar energy conversion, this may consist of a completely new design in energy capture or new principles of materials science. While the spirit of discipline gives us conventional science with gradual advance, the spirit of rebellion often leads to failure because it is more difficult to find better alternatives to existing principles and practice. Occasionally, however, it results in big advances, breaking out of the usual mode of thinking to new and better ways. Eventually, the two spirits come together to be reconciled and form what are called paradigm changes, which are better explanations of nature and yet are based on basic principles of science. Major scientific discoveries often result from such a spirit of rebellion, for example, the change from classical Newtonian physics to modern physics based on relativity and quantum principles, which completely altered our view of the nature of space, time, and matter. Other paradigm changes in science occurred, not through outright rebellion in thinking, but through new views following major discoveries, such as new theories of chemical bonding following the discovery of the electron and a new explanation of genetic inheritance following the discovery of deoxyribonucleic acid (DNA).

**Figure 4.2** Spirits of discipline and rebellion.

The tension between the two spirits, and sometimes their eventual reconciliation, exists not only in science but also in many other domains. For example, in the social domain, up until the beginning of the past century, conventional thinking backed by laws in many countries did not allow women to exert the right to vote. Only active protests by a few women followed first by suppression and then later by extensive soul searching of the societies at large resulted in changes in public thinking, which led to a paradigm change with regard to women's right to vote. The spirit of rebellion was reconciled with the spirit of conformity and discipline, giving rise to a new norm. Today we are still in the middle of paradigm changes with regard to issues concerning gender, race, color, and human rights, among others. At the family and individual levels, we also see frequent conflicts between the two spirits, as when decisions have to be made between following the usual established path and following an unusual one, which may prove disastrous or phenomenally successful. Reason, caution, and a sense of adventure all play a part. Scientific thinking can help in navigating the course of action, since it is familiar with the spirits of both discipline and rebellion, although problems in real life are usually complicated by emotional and irrational factors.

A word of caution is necessary here. Many people have argued in favor of certain actions or solutions ostensibly because they are "backed by scientific evidence." While some assertions may be true, many are complicated by the fact that the so-called scientific evidence is oversimplified, circumstantial, or often just untrue. Should we eat less fat or more fat? Are genetically modified (GM) crops bad for our health or for the environment? Did ancient people communicate with extraterrestrial beings? Science has been called as a crucial witness in many controversial cases. It is advisable not to believe in hearsay, even in the name of science, and to examine evidence more closely and with an open mind. The famous sermon of the Buddha, the *Kalama Sutra* (Box 4.2), can help form guiding principles in this case.

**Box 4.2** The *Kalama Sutra* as Advice for Prudential Thinking

The *Kalama Sutra* (also called *Kalama Sutta*) is a sermon given by the Buddha to the Kalamas clan, which asked for his advice, as many holy men had passed by and had given many conflicting advices critical of one another. The Buddha told them not to believe any teachings just because they are claimed to be true. He listed ten specific sources, which must be rigorously examined:

- Repeated hearing
- Tradition
- Rumor
- Scriptures or official text
- Surmise
- Dogmatism
- Specious reasoning or common sense
- One's own bias

- Experts
- Authorities or one's own teacher

He finally told the Kalamas that only when they know that "these things are good; these things are not blamable; these things are praised by the wise; undertaken and observed, these things lead to benefit and happiness," should they accept the teaching. These words of advice are remarkable, considering the prevailing beliefs of the society at that time over two millennia ago. They are certainly relevant in our time, with so many sources of information around, each claiming authenticity and veracity. The sermon gives a message akin to the spirit of science as we know today.

## 4.3 Cultivating Imagination and Creativity

Can imagination and creativity be cultivated? People have innate differences in their propensity and ability for imagination and creativity. Some people have natural gifts for conceiving ideas, arts, music, business plans, and explanations of natural phenomena. They can think of things about which other people do not have any ideas, see beyond what other people see, and in their minds go where no one has ever dreamed of. Geniuses, as exceptionally smart people are called, tend to be exceptionally imaginative and creative. They can show their smartness, imagination, and creativity, even with little training in their subject of interest. Wolfgang Amadeus Mozart could compose music and perform by the age of five and went on to compose more than 600 works before he died at a young age of 36. Srinivasa Ramanujan, an Indian mathematician who also died young, could solve many difficult mathematical problems at a young age with almost no formal training. Some, such as Alexander Borodin, were geniuses in more than one area. Borodin was a famous chemist as well as a well-known composer. With these examples, we may be tempted to think that geniuses are born with their gifts and that smartness, imagination, and creativity cannot be cultivated. However, many other capable people achieved their success relatively late in life or after having overcome difficulties in childhood. Charles Darwin published his theory of evolution when he was already 50. Although success came at earlier ages, Thomas Alva Edison had difficulties at school and was home-schooled by his mother and Albert Einstein suffered from speech difficulty as a young child. Many people are "late bloomers," who eventually develop to their full potential later in life. Many other examples of people who went on to develop their full potential through their imagination, creativity, and innovation show that these capabilities can be cultivated. The success of such development to full potential depends on the environment at home, at school, and in society in general.

Everyone has imagination and creativity to a certain extent. Like the senses of curiosity and wonder, children are imaginative and creative, but these qualities weaken as they grow older. However, good education, suitable upbringing, and a conducive environment can help retain or even enhance imagination and creativity into later life, producing results in work and other contributions to society. Such outcomes are possible for everyone, not only for the born geniuses. Children should be encouraged to think and dream and to explore and follow their own interests and should be given guidance to develop further without too much premature instruction. Ken Robinson [15] defines imagination as "the power to see beyond the present moment and our immediate environment. In imagination we can bring to mind things that are not present to our senses . . . We may not be able to predict the future but we can help to shape it." In cultivating imagination, therefore, children or older people should be allowed to think or do things out of the box. He distinguished imagination from creativity, in the sense that while imagination is free and may not lead to any actual outcome, creativity is the application of imagination to a concrete conclusion. Creativity is thus applied imagination, the process of bringing imagination to the real world as an art object, a music composition, a new business model, or a new theory of the universe.

As we saw in Chapter 3, imagination and creativity are important ingredients but only parts of innovation. Many skills are needed for innovation. Importantly, modern innovation normally requires good background knowledge of the subject of the innovation, including science, technology, engineering, and mathematics. While the power to think out of the box and the spirit of rebellion are important for innovation, the spirit of discipline and the power to be able to master existing knowledge are also necessary ingredients. In recognizing the power of knowledge, we should take note of what Einstein said [16]:

> Imagination is more important than knowledge. For knowledge is limited, whereas imagination embraces the entire world, stimulating progress, giving birth to evolution. It is, strictly speaking, a real factor in scientific research.

We can explore the relationship between imagination and knowledge further. An analogy can be made to the relationship between syrup and sugar. Imagination is like syrup, with dissolved sugar as the ingredient. The sugar molecules float randomly among water molecules, akin to dreams, questions, and conjectures as yet with no definite direction. When the condition is right, such as when the concentration is high enough, or when the temperature is low enough, small sugar crystals form by aggregation of the sugar molecules. The crystals grow bigger, with a beautiful regular structure comparable to systematic knowledge. In fact, the essence of knowledge is derived from randomly floating imagination. In reverse, we can make syrup by dissolving sugar in water and heating it up, similar to infusion of old knowledge into the pool so as to create new imagination, which can later form new knowledge in an ever-expanding manner. Therefore, imagination can be cultivated, and science education should be an important part of this. Science education should be aimed at creating imagination more than learning what other people have already found out. However, science education that emphasizes imagination should also stress the principles and logic. The syrup solution of imagination should not be so dilute as not to allow crystallization of the sugar of knowledge.

In conclusion, imagination goes beyond the limit of knowledge, allowing us to think and dream of new possibilities. Creativity is the shaping of imagination into a concrete outcome, and innovation is the application of imagination and creativity in combination with a number of skills and knowledge to make new and useful products or processes. Imagination and creativity can be cultivated, especially from childhood, through encouragement and guidance from the teachers, the family, and the general atmosphere and attitude of society.

## 4.4 Cultivating Psychomotor Skills

Psychomotor skills are skills of movement and action controlled by mental activity. They include skills of using the hands, feet, and other parts of the body to do various tasks ranging from delicate ones, such as artistic movements, to those requiring physical exertion, such as sport activities. They are the results of coordinated activities of the cognitive function and physical movement, or brain–body coordination. Like thinking activities, people are born with various natural gifts for these skills. Some are able to do delicate handiwork and some to play sports with ease, while others find it awkward to perform even simple manipulations. Like thinking, psychomotor skills can be cultivated from early childhood, the earlier the better in most cases, provided the bones and muscles are mature enough, as can be seen from training of athletes and musicians, for example. Advanced skills in hand use are special abilities of humans, as no other animals have such dexterity provided by special anatomical features and neural machinery. Even apes, which have hands with somewhat similar anatomical features, cannot command the wide-ranging and delicate use of hands as humans can. Evolution of the hands in humans was accompanied by important changes in the brain, changes in facial features such as relocation of the eyes from the sides to the front of the face, and adoption of the bipedal erect posture, which allowed the hands to do various tasks more freely.

A large part of our education is devoted to developing skills in using our hands and other parts of the body for various tasks and to achieving physical fitness. Music playing, drawing, dancing, and sports are some of the educational activities that help us develop psychomotor skills. Science study also helps us develop psychomotor skills, since it is based on experiments and observation, both of which require such skills. We acquire such skills in practical classes as well as in other out-of-classroom activities, such as using the computer to do various tasks and playing with science-related toys such as Rubik's Cube. Acquiring special psychomotor skills is an important aspect of some science-related professions, such as surgery and scientific drawing (Fig. 4.3) and some branches of medicine and engineering that require such skills in addition to academic ability. In the near future, it will be increasingly important to acquire psychomotor skills, not only for direct operations, but also for control of machine-operated manipulations, such as distant, robot-assisted operations. Skills in computer game playing and other machine-related skills will become more important, and parents should allow or even encourage their children to develop such skills so long as a good balance is achieved between playing and studying.

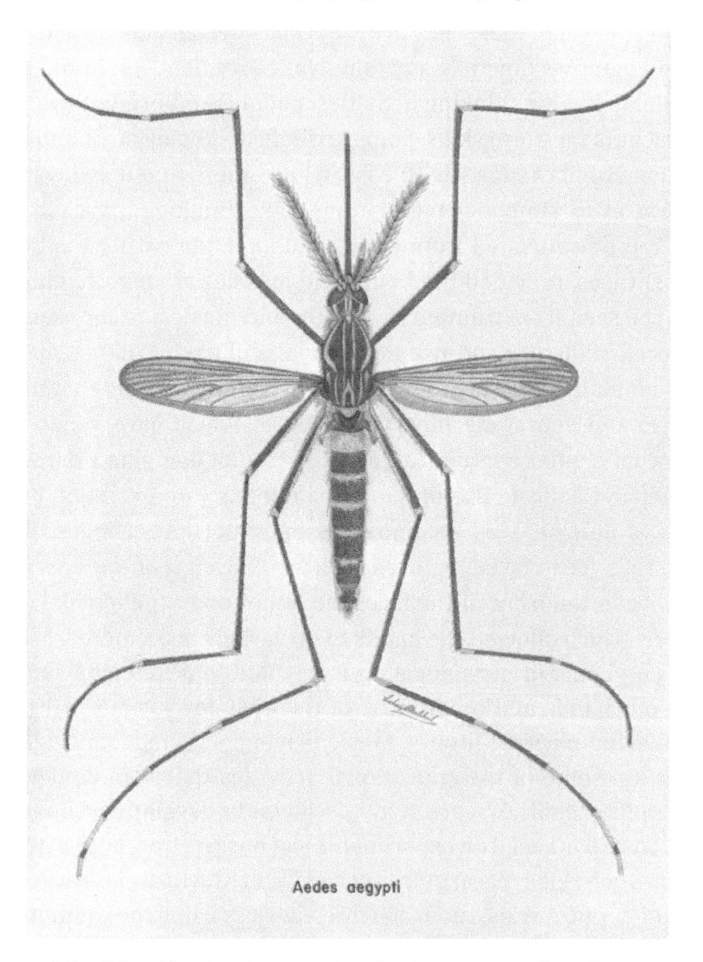


**Figure 4.3** Scientific drawing requires both artistic skill and accuracy in observation. It can depict subjects such as an insect in detail not easily conveyed in photographs. This picture of the mosquito *Aedes (Stegomyia) aegypti,* which causes diseases such as Zika fever, yellow fever, and dengue hemorrhagic fever, was drawn by Vichai Malikul, Department of Entomology, Smithsonian Institution. Picture courtesy of the artist.

## 4.5 Cultivating Passion

We all have our passions, some of which we harbored since childhood, like singing, keeping favorite pets, or following the night sky. Some passions start relatively late, perhaps for something we got to know, appreciate, or were able to develop needed skills for later in life, like cooking, photography, or remote-controlled gadgets. Some passions lead to lifelong engagements or commitments, in raising families and choosing jobs, while some fade away as we grow older. There is evidence of the role of many brain regions—the passion network—in sending neurotransmitters and other brain chemicals when a person falls into deep passion like love. The chemicals create sensations of attraction, arousal, pleasure, and obsession. Presumably, a passion for things other than love can also trigger such signals in the brain.

Passion for many things can be traced to, or linked with, the spirit of science rooted in curiosity, a sense of wonder, and an eagerness for exploration. Some people have a passion for pure science subjects like physics and biology. These people are, however, relatively rare and are sometimes discouraged by friends who call them nerds. Many more people have a passion for things or issues that are related to the spirit of science, like exotic animals and plants, nature trekking, robotics, car racing, or designing and building of things with Lego blocks. These can grow deeper, leading them to develop more knowledge and skills concerning the objects or subjects of their passion. These passions can be encouraged and guided, as appropriate, by teachers, family members, and friends. Passions can be cultivated along specific lines so as to achieve expertise in handling pet animals or manipulating robots. They can also form the starting points for deeper or broader passions in animal biology or robotics. It is up to the circle of people who are close, the environment, and most importantly the ones harboring the passions themselves to decide the courses of action.

A passion for science can grow by itself in some people and can be cultivated to grow in others. Many people are inspired by their teachers, while others are inspired by fascinating accounts they heard in childhood. Some develop the passion gradually, perhaps as they mature and are able to see bigger pictures after their early passion with science-related subjects. The passion tends to grow following success in doing the things they love or in knowing more about the subjects of their passion. Provided that the specific subjects of the passion are not morally or legally questionable, we should follow our passion, in science or in other subjects, and should encourage others to follow theirs. Passion is the important motivator for our action and gives us the power to accomplish our goals, to do things with our hearts together with our heads and hands. However, we need to consider the long-term consequences of following a passion and not always follow it or urge others to do so blindly. We need to exercise our reason as well as cultivate our passion, as Rudyard Kipling says [17]:

> Your reason and your passion are the rudder and the sails of your seafaring soul. If either your sails or our rudder be broken, you can but toss and drift, or else be held at a standstill in mid-seas . . .

## 4.6 Which Side of the Brain to Cultivate

Studies on the human brain show that different functions are located in different parts of the brain. In brief, these functions are distributed in both the left hemisphere and the right hemisphere, which have extensive connections so that the brain functions as a coordinated whole and not as independent parts. The left hemisphere controls the right side of the body and functions in using and understanding language, logic, and mathematics. The right hemisphere controls the left side of the body and functions in spatial coordination, nonverbal information, and emotion. Some people simplify the left hemisphere as the logical, methodological, and analytical side and the right hemisphere as the artistic and creative side of the brain. This has been shown to be an oversimplification, since we use both sides of the brain in a coordinated manner. In learning and practicing science, for example, we have already seen the importance of imagination and creativity as well as reason and logic. We are therefore using both sides of the brain at the same time. Similarly, people who are doing artistic or other creative works also use various functions of the brain together. Sometimes we need to concentrate on the use of the analytical and logical functions, while at other times, we stress the creative and artistic functions, but overall both sides of the brain are called into use.

It is clear, therefore, that we need to cultivate the functions of both brain hemispheres, not of one or the other. It is true that for some tasks, we need to call on some specific functions more than others. In general, we need to cultivate all the functions of the brain as well as their coordination and connections, with emphasis on the functions that are needed especially for our specific tasks. For example, as technical analysts or software engineers, we may have a special need to cultivate the analytical, logical, and mathematical sides of our brain. Being able to cultivate all the functions of the brain appropriately will result in a balanced brain and a balanced mind yet with a special ability for our specific tasks.

## 4.7 Cultivating Multiple Intelligences

What we know about the various functions of the brain leads to a better understanding about intelligence. Some people such as Howard Gardner [18] proposed that instead of the concept of intelligence as a single attribute, as traditionally assumed, we should think of multiple aspects of intelligence. People may be intelligent in some aspects but not others, and a single measure such as IQ is not adequate to describe their intelligence. Gardner proposed eight intelligences to account for different abilities of people in different things: linguistic intelligence (for words), logical-mathematical intelligence (for numbers and reasoning), spatial intelligence (for space and visualization), bodily kinesthetic intelligence (for physical activity), musical intelligence (for music), interpersonal intelligence (for social interaction), intrapersonal intelligence (for reflection), and naturalist intelligence (for nature and the environment). Recognition of multiple intelligences has far-reaching implications on education and human development in general. In traditional education, people who have high intelligence in certain areas but not in others may not be recognized as being intelligent and may not get opportunities to develop to their full potential. In recognizing that people have multiple intelligences, which may be revealed even when they are young, they will get better opportunities to develop their special intelligences to the full. This is in line with the need for various 21st-century skills, which, as we discussed earlier, comprise skills for various types of activity both in life and in work. People with certain exceptional intelligences will be suited to life and work of certain types and together make up a balanced society with various needed skills as a whole. For science in particular, development of people with complementary multiple intelligences who make up the community of scientists, engineers, and related personnel will lead to a stronger spirit from which the sparks of innovation and sustainable development can be better generated.

## Chapter 5

# Creating Sparks from the Spirit

*Creativity is a type of learning process where the teacher and the pupil are located in the same individual.*
—Arthur Koestler, *Drinkers of Infinity: Essays, 1955–1967* (1968)

The spirit of science enables us to create sparks, namely discoveries, innovations, and subsequent development processes. The sparks are not automatically created from the spirit but need knowledge and skills and, more importantly, creativity and inventiveness. Knowledge can be learned, while creativity and inventiveness arise from the ability to go further into action, such as building new tools or solving new problems. Like the ability to think, the ability to create and invent is innate in everyone and can be cultivated further by upbringing and education that encourage exploration of new ideas and experimentation, not just memorization and conformity to established practice. Inspiration; learning from past experiences of others; trial-and-error practice, extending into new ventures with science and technology as a knowledge base; logical planning; and the ability to integrate different aspects of related tasks, all contribute to creativity and inventiveness. The capability for discovery and innovation is the outcome of creativity and inventiveness mixed with the application of knowledge and other skills. A major role of education, in addition to equipping the learner with knowledge and skills, is to enhance creativity and inventiveness, increasing

*Sparks from the Spirit: From Science to Innovation, Development, and Sustainability*
Yongyuth Yuthavong

ISBN 978-981-4774-57-4 (Hardcover), 978-1-315-14599-0 (eBook)
www.panstanford.com

the potential for successfully igniting the sparks of discovery and innovation.

## 5.1 Making the Connection between the Spirit and the Sparks

The spirit of science leads us not only to learn what others before us have done and accomplished in discovering new knowledge and bringing about innovations but also to imagine further about the nature of the world around us and possibilities of bringing about our own innovations. The spirit leads to sparks of discovery and innovation in scientists, technologists, entrepreneurs, and others, with great benefits for society. But can we also be among those who generate discovery and innovation by ourselves? Can we create sparks from the spirit?

The answer is that it is possible to do so. However, it requires more than just knowledge, skills, and familiarity with the fields of expertise. These are important, but they form only the basic part of the requirements. The situation is similar to painting, where we can learn and know much about the theory of painting, nature of light, properties of paints, beauty, and art and yet not become painters until we actually paint and the pictures actually come out beautiful. So what do we need to go from knowledge and skills about a subject to discovery and innovation? The first thing we need after equipping ourselves with basic knowledge and skills is belief in our own capability. This comes from realistic self-appraisal of our present status and future potential. We cannot all be Albert Einstein or Thomas Alva Edison, but most of us can find or do something new and useful. There is a whole range of potential new knowledge and inventions, where we can become part of their origins and development. We can all contribute to generating sparks, big or small, from the spirit of science. However, we must first equip ourselves with basic knowledge and skills about the subject to which we want to contribute, and we must be aware of what has been going on so that we do not "reinvent the wheel." It is amazing that so many things that we think are new have, in fact, been explored, discovered, or invented by other people. Equipping ourselves with basic knowledge and being aware of what has been going on help us not to try to ignite the sparks that have already been ignited by others before us.

Although most of us do not expect to discover something of great importance in science or invent something of great use and novelty, we can still produce small sparks from our endeavors that we can be proud of. A better and more elegant solution to a mathematical problem than that given by the standard solution, a better way to understand a scientific rule than that offered by our teacher, a better way to play Rubik's Cube, all these do not give us fame or fortune but make us proud to be able to discover or make something new by ourselves. However, it is much harder to discover or invent something of significant value to the society at large. Normally this requires long study in our fields of endeavor so that we know the details and status of the subjects. We can also expect that although we may discover or invent something new to our knowledge, it is quite possible that other people have also independently done so. Indeed, multiple independent discoveries or inventions are quite common throughout history (Box 5.1). Moreover, because of the complexity of the subjects, we normally would expect to be part of the team involved and not the sole discoverer or inventor. Sole inventors or discoverers, common in the past centuries, are now quite rare.

**Box 5.1** Independent Discoveries and Inventions

Independent discoveries and inventions occur quite often. Sometimes the discoveries and inventions are made entirely independently, without knowledge of other persons working on the same or similar ideas. In many cases, the independent discoverers or inventors have ideas from the same sources, such as previous works by others on the subjects of their interest, or may even have communicated with one another. It is difficult to decide just by allegations from the purported discoverers or inventors as to the priority of their works over competing claims.

A famous controversy concerning the priority of invention is that between Sir Isaac Newton and Gottfried Leibniz over the development of modern calculus (Fig. 5.1). Mathematics of changes and motions, which are embodied in modern calculus, had been studied by Greek and Indian philosophers long before the time of Newton and Leibniz. Newton began his work on the subject in 1666 and called his method, which would form the subject areas of differential and integral calculus (concerning rates of change and summation of quantities, respectively), the method of "fluxions and fluents." However, he did not publicize his findings, and around 1673, Leibniz developed his form of calculus independently and published his work on differential and integral calculus in 1684 and 1686, respectively. In 1687, Newton published his monumental work *Principia Mathematica,* in which he described his method of calculus formally. Newton accused Leibniz of plagiarizing his work, and a bitter controversy ensued. In 1715, the Royal Society proclaimed Newton as the true founder of calculus. As time passed, however, the world acknowledged the independence of Leibniz and the value of his methods of calculus, and both Newton and Leibniz are now proclaimed co-inventors of the subject.

Charles Darwin is accredited with the discovery of the theory of evolution, concerning the struggle for survival of species and random changes that would decide their fate. He obtained evidence for his discovery from his voyage on the exploration ship *HMS Beagle,* which stopped at various locations in the world. Independently, Alfred Russel Wallace accumulated evidence for an evolutionary theory from his voyage to the Far East, especially Southeast Asia. The theory of evolution can be said to be co-discovered by Darwin and Wallace.

Other cases of independent discoveries and inventions have occurred all through history. Modern examples can be cited, from the discovery of the catalytic properties of ribonucleic acid (RNA), residing in what are termed "ribozymes," to the discovery of methods of editing the genome. The invention of driverless, automated cars by a number of companies can also be regarded as independent, although it is a composite invention consisting of many contributing parts, each with its own history of development.

We can perhaps explain why there have been so many independent discoveries and inventions from the fact that development of various fields of science and technology often comes to stages where leapfrogging is likely. Disruptive discoveries or inventions, which represent jumps in knowledge or the level of innovation, occur sporadically when the situations allow ingenious people to make such jumps. These innovators depend on previous background knowledge and advances in their fields, but it is due to their imagination and creativity that they can make the leaps to produce substantial discoveries and radically new inventions. History has shown that such imagination and creativity often arise in many gifted people at the same time.

**Figure 5.1** Who invented calculus? Isaac Newton and Gottfried Leibniz both claimed to have invented calculus.

## 5.2 The Importance of Scientific Knowledge

Scientific knowledge is important for making discoveries and bringing about innovations. In the old days, inventions could be made from observation, trial, and error without much reliance on science and technology. In the past couple of centuries, however, such inventions and discoveries based on simple observation, trial, and error were exhausted, while the knowledge base became more important. It was pointed out [20], for example, that while fabrication of a reaping machine from metals and wood would be conceivable with trial and error, invention of the radio would be impossible on a trial-and-error basis alone without prior knowledge about physics of semiconductors and other scientific areas. Scientific knowledge has become the basis of technology. In some cases, technologies arise like offshoots of scientific knowledge. Medical imaging from physics of radiation, plastic technology from chemistry of polymers, and new pharmaceuticals from biosciences are examples of such offshoots. Trial and error play a part in some activities, such as in the early development of the steam engine by Thomas Newcomen and James Watt, but they were not completely random activities. Instead, they were based on what was then known about the science of heat and steam. Nowadays, selective plant breeding, screening of drugs from stocks of chemicals, and selection of promising products from large numbers of candidates still rely partly on a trial-and-error basis. However, even in these examples, science has facilitated and enhanced the potential outcomes. In plant breeding, molecular markers (indicators for the required traits) are used to guide the search, so it is not completely random but is based on testing of some important desired traits. In screening of compounds as potential drugs, high-throughput assays based on automated procedures help researchers discover hits that can lead further to development of new drugs. In other cases, science forms the cornerstone of technologies that are also based on other elements. Food technology has the sciences of nutrition and agriculture as its foundation, among other factors, including markets, culture, and social values. Sports technology is a derivative of physiology and psychology of exercises, mixed with many other elements, including community values, curriculum setting, and organization of sports events. Even innovations that seem to be socially based rely now on science as a main enabler. Social networks, which were originally based on direct person-to-person contact, now rely considerably on the Internet and other technologies, in turn based on communication and information sciences, with combined powers of speed and outreach. All of these examples underscore the importance of knowledge, especially scientific and technological knowledge, in innovations and discoveries.

## 5.3 "Imagination Is More Important Than Knowledge"

Important as knowledge is, we recognize the words of Einstein [16] that "imagination is more important than knowledge," which we discussed briefly in Chapter 4. Imagination is important for both scientists and inventors in the creation of their works, as can be seen from the examples of the works of eminent scientists like Einstein and eminent inventors like Leonardo da Vinci. Imagination goes outside the boundaries of knowledge and often results in discoveries and inventions. By freeing oneself from the boundaries of knowledge, the imaginer can become engaged in new concepts that could be irrational and mystical and are expressed not as science and engineering works but as other creative works, including paintings, sculptures, architecture, music, and poetry.

Imagination in science and technology is different from imagination in other areas in one important aspect. While imagination in other areas can lead directly to the creation of works of art or other output, imagination in science and technology has to pass some crucial tests. The fact that science and technology are based on rational thinking and accumulated knowledge means that imagination is the beginning of a long process of elaborating on new concepts. These concepts may seem quite wild and impractical at first, and reconciliation with established scientific principles or current technological capabilities is needed. For example, many people have imagined machines that can go on working without input of energy, perpetual motion machines as they are called, but such imagination needs to be tempered by the principle of conservation of energy, which states that energy can be neither created nor destroyed. Such imagination is therefore futile and cannot lead to any new devices. Current technological capabilities also limit how far imagination can go toward reality. Da Vinci was a prolific inventor as well as an artistic genius. He imagined many types of flying machines, a drawing of which is shown in Fig. 5.2, well before they became a reality a few hundred years later. This requirement for checking of imagination with the current scientific knowledge or state of the technology is an additional element that makes it different from imagination in other areas. In the case of da Vinci, we are amazed at his leap of imagination well before it became reality. In the case of perpetual motion machines, they are still impossible. Energy cannot be created anew out of nothing but must come from some sources, such as other forms of energy or from transformation of mass as in nuclear processes.

**Figure 5.2** Design of a flying machine operated by a human by Leonardo da Vinci (circa 1485). Da Vinci imagined a number of flying machines well before the science and engineering of flight were mature enough to make such machines possible in the early part of the past century. From www.leonardodavinci.net.

Einstein also said that he believed in intuition and inspiration, which made him at times feel certain that he was right, while not knowing the reason. Intuition is the ability to reach a new concept without reasoning, and inspiration is a force or influence that moves a person to do or create something. These are the important elements we need beyond knowledge in making new discoveries and bringing about innovations. This is clearly understandable since knowledge consists only of what is already known and not what is new. Imagination is conceiving of what has not been known to exist before, often with the help of intuition as the seed and ramification of the idea for making a leap into the unknown. It is also fueled by inspiration, by persons whom we admire, or from our own inner fire. It can arise seemingly out of nowhere, like a flash without warning, from a conversation, a moment of reflection, or a dream. Often, it results from long obsession or contemplation about a new concept or design that would introduce new knowledge or invention. The successful imaginer can be someone who has been working in the field for a long time, hence knowing the background and stumbling blocks and being able to sense new directions. On the other hand, the successful imaginer can be new to the field and hence be able to conceive of fresh visions and new ideas without being saddled with ones already preconceived by others.

A story was told of the discovery of the structure of benzene, as a ring of six carbons, by August Kekulé [21]. He had been thinking about how atoms of carbon could form so many compounds. In a daydream on a bus ride in London, he saw carbon atoms joining in a giddy dance, giving him the idea of how they could be linked with one another in a chain. In another dream he saw the dancing strings of carbon like a snake coiling in a circle and swallowing itself (Fig. 5.3). Self-swallowing snakes, known as ouroboros, have been represented in religions and myths since ancient times, symbolizing reflexivity and the cyclical nature of things. These two dreams of Kekulé were likely the outcome of his continued imagination about structures of carbon compounds. Another example of the role of imagination was from the discovery of the structure of deoxyribonucleic acid (DNA) as a double helix. James Watson and Francis Crick were relatively new to the field of DNA, but were quick to learn about previous studies into its structure. Their discovery owed partly through their imagination of the structure as a spiral staircase, with the base pairs forming the steps [22].

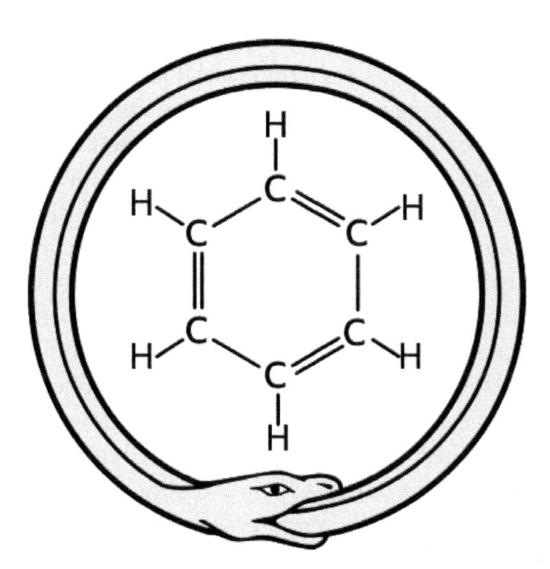


**Figure 5.3** August Kekulé's dream of a snake following itself was the inspiration for his discovery of the structure of benzene. Picture by Haltopub from Wikimedia Commons under the Creative Commons Attribution-Share Alike 3.0 Unported license (https://creativecommons.org/licenses/by-sa/3.0/deed.en).

These stories offer anecdotes of the role of imagination in discovery and invention. Indeed, imagination is behind every discovery and invention, since it is needed for creating something new. In the case of discovery, imagination leads to new knowledge. In the case of inventions, imagination leads to new, useful devices. However, we have noted that imagination can lead to many outcomes, as science, engineering, and other creative works such as art forms. Important though imagination may be, other elements are also needed to go on from imagination to discoveries and inventions.

## 5.4 From Imagination to Creativity and Inventiveness

Imagination is the source of creativity. When we want to create something new, we must first imagine it. However, imagination alone is not enough since it only stays in our minds and does not lead to a discernible outcome for the outside world. We can imagine something far out of our own normal experience, such as traveling in time back to the past, visiting another planet, or being invisible. However, without the existence of physical laws or tools to achieve the objects of our imagination, it will just remain wishful thinking. This does not mean that we cannot create something out of these imaginations. Science fiction, films, and other creative works have been made on the basis of imagination of things or situations that are far from our everyday experience. Some of the works based on these wild imaginations have become, or are close to, reality. Jules Verne's *Twenty Thousand Leagues under the Sea*, written around 1870, anticipated the many capabilities of today's submarines. Isaac Asimov coined the term "robotics" and foresaw a number of advanced robots now appearing in reality. Space travel, previously seen as largely fantasy when Stanley Kubrick and Arthur C. Clarke made the film *2001: A Space Odyssey*, is now reality. That film portrayed a rebellion by the on-board computer Hal, foreseeing a contentious issue of obedience of artificial intelligence to human creators. Other works of fiction such as *1984* by George Orwell, depict a world where the thoughts of people are controlled by Big Brother and free thinking is a crime. *Brave New World* by Aldous Huxley describes a future world where people are raised in hatcheries and predestined into ranked castes. These are works based on imagination in science or imagination about the future where science plays a big role. They are not physical products or processes but creations that not only are entertaining but also make us ponder the societal and ethical implications of scientific advances.

Creativity and inventiveness leading to new products and processes in science and technology require imagination as the initiator. However, in going on to create or invent something real, the imagination has to be checked with reality. A reality check is an important step, since all too often the imagination is not unique and has already been transformed into a product by someone else. The check must be made if we need to create something new, as either an invention or a discovery. For example, the three basic requirements for patenting a new product or process are that it must be new, useful, and non-obvious. The basic yardstick for a scientific discovery is that it must be new, in the sense that it has not been published or known to the public through any media before, and must add to the stock of knowledge in the field of the claimed discovery. The imagination that can go on to produce sparks of creativity and inventiveness must contain originality, something that no other person has conceived of before, or at least not in exactly the same way. This means that the imaginer must study the state of the art in the field of research and development involved and must make a thorough effort to find out whether other people have also imagined similar concepts and have already gone on to create something new from these concepts. If not, the way is clear for the imaginer to move on to the next step. Quite often, however, many people will have the same imagination or new ideas about a certain field independently as the field matures and opens up new possibilities, and the burden of proof of originality can become a complex legal problem. For example, new ways to edit the genome, to be able to correct genetic defects or produce organisms with desired properties, have recently been published and patented by a number of research groups. Many other cases of contending priorities in discoveries or inventions have been fought, both in public and in court. A happier and more productive course of events is cooperation among the different people working on the same, or similar, problems. Open innovation or networking among researchers has become a widespread mode of working, as discussed in Chapter 2.

As noted earlier (Section 4.3), creativity is a step further from imagination. As Ken Robinson observed [15], "Creativity involves putting your imagination to work. In a sense, creativity is applied imagination." To put imagination to work, we require realistic conceptualization, design, and a work plan concerning our foreseen innovation. These processes bring our imagination to confront with the reality of the state of the art and science concerning the object of our imagination. Robinson went on further to say, "Innovation is the process of putting new ideas into practice. Innovation is applied creativity." In going on from imagination to creativity, and further to innovation or inventiveness, we need to be armed with the necessary tools. Such tools are basically in the form of multiple intelligences, which we discussed in Chapter 4. We recall what Howard Gardner [18] said about multiple intelligences: "People have different intelligences for one or more different things, including linguistic, logical, spatial, physical, musical, interpersonal, intrapersonal and naturalist intelligence." In going on from imagination to creativity and inventiveness, we need to apply some of these intelligences, either on our own or relying on group effort. For example, we may imagine having a new computer game on, say, climate change. We can imagine all sorts of scenarios where players can act to mitigate or worsen the climate situation. After the phase of imagination, we must think about creative ways to play the game, taking into account the real situation on climate and its science. We need to have a conceptual design of the game, including the number of players and how they can play, as individuals or as groups. We need to go into the issue of developing or adapting suitable software and choosing devices. Many more considerations are needed, in greater and greater detail, which require our multiple intelligences. It is quite likely that at some point we will need to find a team to work on the project, or go to professional game developers to sell our ideas. Other cases of innovation may be made by one or a few persons, but the reality of today is that most scientific and technological innovations need to be developed by teams of workers with different expertise. Still, the person who first conceives of the idea, the one with the first imagination, is the most important one, for without this imagination the whole creation would not exist.

## 5.5 Learning Creativity and Inventiveness

Can creativity and inventiveness be taught or learned either from the classroom, upbringing, or the social milieu? Inasmuch as creativity and inventiveness spring from imagination, and imagination is an individual trait, it appears at first sight that they cannot be taught either in class or outside. However, we should note that creativity and inventiveness tend to be found in groups of closely associated people or students of the same teachers or descendants of creative and inventive people themselves. While it can be argued that descendants of such people have inherited the traits for creativity and inventiveness, we must consider the likelihood that factors other than genes are also important. Studying creative writers like Mary Shelley, the author of *Frankenstein*, one of the earliest science fiction works, Cody Delistraty [23] found that the common traits among highly creative individuals are openness to experience and an ability to make connections between seemingly separate issues and to produce something new out of these ingredients. In the case of *Frankenstein*, a combination of ghost and horror stories, the then-new findings of the effects of electricity on animal organs, her dreams and nightmares, and the influence of writers like Lord Byron, whom she was staying with, all had an effect on Shelley's conception and writing of the story. In the conception and discovery of real scientific knowledge, Richard Feynman, a brilliant physicist who won the Nobel Prize, stressed the importance of following up on curiosity about anything, having playfulness and pleasure in finding things out, having the ability to delve deeply into a subject and yet to make it simple, and in knowing something really well [24]. For Steve Jobs, creativity is about connecting things: "Creativity is just connecting things. When you ask creative people how they did something, they feel a little guilty because they didn't really do it, they just saw something. It seemed obvious to them after a while. That's because they were able to connect experiences they've had and synthesize new things. And the reason they were able to do that was that they've had more experiences or they have thought more about their experiences than other people" [25]. Jobs was influenced by his earlier trip to India, when he was struck by the observation that people in India did not respect intelligence as much as instinct. These few stories tell us that although creativity may be an inborn character to a certain extent, it can be learned from examples around us, from people we are associated with, and from observing the ways other people create.

Examples from people we admire can give us inspiration to become creative. Teachers who first introduced us to fascinating subjects in class, friends who thought of something interesting and adventurous, creative people whose stories we read or see on TV or had opportunities to acquaint ourselves with, and, when we reach the stage of higher learning, mentors and coworkers who worked together with us in research and engineering projects, all these people can give us examples from which we can draw inspiration. Take the subject of diversity in nature as an example of an inspiring topic. A teacher can inspire us when we are helped not only to understand natural diversity but also to be creative in devising ways to study and conserve nature. Friends can inspire one another when they participate in a discussion on a subject and create something together or become engaged in friendly competitions. We can be inspired by our friends to find creative ways to examine natural diversity worldwide through the Internet. We can learn from, and be inspired by, people who have worked on natural diversity, from Alexander von Humboldt, the naturalist explorer from three centuries ago, to Sir David Attenborough, the naturalist broadcaster of today, and imagine further from their examples how we might ourselves become creative explorers. In all areas of investigation and innovation, we can learn from, and become inspired by, the imagination and creation of various people whom we admire.

Parents and the immediate family have a large role to play in encouraging the creativity and inventiveness of children. In addition to genetic characters from the parents that may influence these traits in children, the environment in the family provides opportunities through daily interactions including play, conversation, and other activities. Reading to children, for example, not only is entertaining but also encourages them to explore various aspects of their universe through imagination. Parents can inspire their children through introducing interesting stories and answering the questions that they pose. Children express their early creativity through such means as drawing, play acting, and playing with toys through imaginative situations. Both in the home and at school, creativity can be encouraged through responses that allow imagination and further exploration. Parents and teachers can guide development of creativity in children by paying attention to their ideas and expressions, such as making or playing with toys and inventions, and helping them to experiment with their ideas. At this stage the product of creativity is not so important; it is the process of creativity that is important and should be encouraged.

## 5.6 From Idea to Practice

In science and technology, creativity and inventiveness lead to new products in the form of knowledge or useful devices. These products are different from the products of creativity in arts, music, or other areas, which are entirely the creations of the originators. Certain aspects of the process of creation are similar, such as following up the initial imagination with action. As we noted earlier, unlike works of art or other creative works that come entirely from the originators, the products of creativity and inventiveness in science and technology need to pass through the hurdle of a reality check, that is, a check whether they are really innovative, useful either as a device or new knowledge, and not already discovered or invented by others. The process of going on from imagination and conception of the idea to creating and realizing the fruits of creativity in terms of discovery and inventions is a long one, involving the use of many types of intelligences and skills, as already outlined (Chapter 4). Here we summarize the main aspects in going on from idea to practice, resulting in the final products of creation.

### 5.6.1 Guidance by Passion, Curiosity, and Inquisitiveness

In addition to a general passion for the subject of interest (Chapter 4), creativity requires a specific passion for achieving the results of our imagination. In some cases, the passion is self-generated, while in other cases, the passion may be derived from, say, the job we are doing. In the latter case, we should go on to develop our own passion, not just treat it as a job assignment. Together with having passion, we need to be intensely curious and inquisitive about various facets of our potential creation. For example, we may have a passion for discovering new drugs, which may be derived from our general passion for chemistry or medicine. In going forward, we need to be aware of the status of the field in which we are interested, say, drugs against infectious diseases, to be curious about various problems, such as drug resistance of infectious microbes, and inquisitive about the ways in which people have attempted to solve these problems.

### 5.6.2 Aiming for a Target and Checking for Novelty

Aiming for an appropriate target of our creation is the most important part of our quest. Aim too high and the work is doomed to failure; aim too low and we are condemned to spend valuable time and effort in unworthy activities. The target should be significant in that it would be something new and/or useful in our field of work and something that we can realistically hope to achieve in a reasonable amount of time, considering the resources available.

Two types of targets can be discerned, incremental or disruptive, like walking up a spiral staircase or jumping from one level to another, respectively (Fig. 5.4). An incremental target is one that adds to the stock of knowledge or inventions, with significant value, but represents no threat to the status of the field. A new drug belonging to a known family of similar drugs is one example of an incremental target. A disruptive target is one that drastically alters the field, for example, the first cell phone in the age of land lines. In science, such a disruptive target is the result of what is called *paradigm shift* by Thomas Kuhn [26], which is something that changes the status of the field significantly. Jumping up the staircase by a level, rather than walking up step by step, is much quicker but bears the risk of falling down! Einstein's relativity theories are examples of such discovery, disrupting the field of classical physics dominated by Newton until then.

Once we have a suitable target in mind, we should go on to check for the novelty of our ideas and target. This is done by checking on patents and publications concerning our field of interest, by going to scientific meetings, and by inquiring among the network of researchers in this area. Often, other people have also had similar ideas and may already have gone on to achieve similar targets. This does not mean that we abandon our own ideas and target, unless it is clear that we have lost the priority for discovery or invention, but it means that we should make adjustments to our original ideas and target so as to make them still worthwhile.

**Figure 5.4** The spiral staircase of the Vatican Museum. Discoveries and innovations can be compared to going up a spiral staircase. Incremental discoveries and innovations are like walking up the staircase step by step. Much rarer disruptive discoveries and innovations are like jumping across from one level to another, which are riskier but when successful result in a paradigm shift or drastically new inventions. Picture by Na Dol Vatanatham.

### 5.6.3 Planning and Execution

Creative works need planning, just like building houses or running businesses. However, the plan should not be rigid and should leave room for adjustment or unexpected development. In most cases of modern research and innovation, the work will need significant financial resources, and it needs to be done by a team in a suitable institution with the necessary equipment and infrastructure. The researcher or inventor needs to be able to write a good research project for funding by appropriate bodies. Often, the availability of funding is the decisive factor for the project, and the researcher or inventor needs to be able to convince the funding agencies of the merit and novelty of the work. The work plan has to be realistic and can go ahead quickly once the resources are found or, in many cases in today's research and innovation, can go ahead even before specific funding is found through support from a related project. The work should be executed, or be under intensive guidance, by the principal investigator or the creator of the project.

### 5.6.4 Tenacity and Hard Work

As Edison said, "Genius is one percent inspiration and ninety-nine percent perspiration." Our ideas and imagination may have been the initiator of our attempted creation, but it is the following hard work that determines the success of such creation. Often, the work goes through difficult phases, including lack of funding and personnel, difficulty of experimentation and making prototypes, and unexpected results and obstacles. Some problems are innate to the nature of the project: for example, in drug discovery, only about 1 in 10,000 new compounds made for such purpose can survive the selection process to warrant commercial production. The investigator therefore needs to be prepared to meet with disappointments, but not to give up easily. In many cases, a less ambitious goal can be attained, such as discovery of the types of compounds that are promising for further development into drugs, rather than discovering specific drugs. Tenacity and hard work in going through with planned research and innovation, after appropriate adjustments in the light of difficulties and unexpected results, are important in deciding the eventual fate of the project.

### 5.6.5 Organization and Networking

The complexity of today's research and innovation sometimes means that no single individual or team is responsible for the final success of a project. We already saw (Chapter 2) that open innovation is now increasingly adopted in many areas, such as early drug development or information technology projects. To achieve the results of our creativity, we usually need organization of our own team and networking with others. There are roughly two types of arrangement, horizontal and vertical. A horizontal arrangement is made within the team or with others in distributing responsibilities, collaborating, and communicating with one another on the objectives and progress of work that are roughly at the same stage of development. A vertical arrangement is made so as to move the work from one stage to the next, say from the research stage to prototype building and further to market production. Confidentiality is an important aspect of such an arrangement and may need to be made as formal documents so that the progress of the work is not disclosed prematurely. Arrangement also has to be made to distribute responsibilities, workloads, and benefits that will later accrue among the teams and team members. Multiple intelligences are needed in these arrangements since they involve various aspects of work. It is preferable for the main creator of the work to possess such multiple intelligences, but he or she still needs collaborators with various skills and intelligences to help with various facets of the work.

### 5.6.6 Going through to the Finishing Line

Starting from imagination, the long process leading to a creative outcome would be futile without the final product of creativity. It is therefore important to go through to the finishing line. There can be more than one finishing line as one success can lead to others down the road, and the process may not be an all-or-none one, but there are important milestones to pass. In scientific discovery, the milestones are publications of the works in recognized scientific journals. In invention, the milestones could be the production of prototypes or final production for the market. Often an important discovery or invention will lead to related or improved ones down the line. The finishing line may therefore be the starting point for further creative works.

The scenario described here is just one of many possible ones in going from imagination to creativity. For a young scientist or inventor, this seemingly long and involved process could be discouraging and deterring. However, everyone, no matter how young or inexperienced, can participate in a creative process, either as a team member or as the main creator. Box 5.2 gives advice to young scientists and inventors who would like to participate in a creative process like research and innovation.

**Box 5.2** Advice to Young Scientists and Inventors on Enhancing Creativity

Creativity can be enhanced, especially when we are young. We can let our imagination fly around the subject that we already have some background of and can sense some room for improvement, such as the need for a new explanation of some natural phenomenon, for a new mechanical device, or for a new computer program. Although we should let our imagination fly as freely as possible, we also need to be aware of the status of the field in which we would like to create something new so as not to spend our time and effort in futile pursuit.

Our attempts at creativity should be guided by the forces of passion, curiosity, and inquisitiveness. It will also be helped tremendously by two other forces, namely the force of knowledge and the force of skills. The force of knowledge comes from our familiarity with the field of our search and related areas, and the force of skills comes from practice in these fields. Young scientists and inventors may need to accumulate these forces with growing experience. We can learn both by ourselves, which is now much easier than before with the availability of online learning contents and other means of learning. We should cover both the depth and the breadth of the subject, going as far as we can, but we should not be bogged down in too much detail, which can blur our vision and imagination. The yardsticks for our achievements are quality and professionalism in learning, not just accumulation of unrelated materials without plans. This process of accumulation of the force of knowledge and skills is valuable, and although we may eventually not end up with the creative achievements originally planned for, we can use this process in pursuing other creative efforts in the future.

In many cases, young investigators start by being students or junior members of teams of scientists and inventors working in specific areas of research and innovation. In such cases, we should try to learn as much as possible from the chief investigator and team members. A specific work assignment may be given, in which we should try our best to become familiarized with the status of the field and the nature of the assigned problems, together with backgrounds of previous attempts. Gradually, we gain enough confidence and expertise to imagine and ask questions on our field of pursuit, in collaboration with the team or on our own, which we should discuss further with the chief investigator and the team. We become more self-reliant, and the problems that were originally given to us become our own problems around which we can creatively search for solutions. We gradually change from a follower to a coworker or even team leader. However, no matter what status we assume, we should always respect our mentors and team members and give them due credit when our own effort bears fruit.

# Chapter 6

# What Is Science For?

> *. . . And science is simply common sense at its best . . .*
>
> —Thomas Henry Huxley
>
> *The Crayfish: An Introduction to the Study of Zoology* (1884)

The purpose of science is mainly to gather knowledge about nature and to obtain benefits from such knowledge. Science is a source of knowledge for daily life—and we use its principles or products routinely—but its base goes both deep and broad to cover much more than just being a sourcebook for routine usage. It enables us to understand the nature of things and helps us prepare for a better future. It continuously opens up new frontiers through research and development. The knowledge is cumulative and self-correcting, owing not to just individuals but to all who contribute to it through discoveries, which together form the big picture of nature. Science is the main source of technology, which gives rise to useful products and processes. Both advanced and developing countries can employ science and technology to enhance their status of development and approach sustainability. Due to its role in acquiring knowledge and giving benefits, science can potentially contribute to the development of humanity and environment in a sustainable manner.

*Sparks from the Spirit: From Science to Innovation, Development, and Sustainability*
Yongyuth Yuthavong

ISBN 978-981-4774-57-4 (Hardcover), 978-1-315-14599-0 (eBook)
www.panstanford.com

## 6.1 The Many Purposes of Science

What is science for? Ask this question, and the answer you get will be varied, depending on the person answering. Ask someone who fails to be impressed by nature and does not want to struggle with mathematics, and you can get the same question thrown back at you in a negative response. Most people would give a neutral answer, somewhat like accepting that science is good for society, although they themselves do not have much to do with it. However, some people think more positively, namely that science is fun, a challenge for the brain, and a source of answers for inquisitiveness. For the people with positive attitudes toward science, we can go deeper into the question, and as we do so, we find that science serves many purposes, some of relevance to some people and other of relevance to other people. Even people with initially negative attitudes can be convinced that some aspects of science serve some purposes to them. We recall that the spirit of science, which chiefly arises from our curiosity, sense of wonder, and eagerness for exploration, remains dormant in all of us, although in many people it has been repressed or neglected as they grow older. Recall that we ask the questions of why, when, where, how, what, and who often in our daily lives and that these questions form the crux of scientific investigation. Therefore the question on what science is for can be answered in a broad sense, in that it is for making us reasonable persons, capable to some extent of solving our own problems through logic and the basic principles or findings of science. Many people enjoy mystery novels, science films and fictions, board games, and other games in general that require strategies and plans. Science uses similar methods as they do in figuring out who the killer is, what the scenarios of the future world are, and how to defeat the opponents. Therefore, at the basic level, the answer to the question of what science is for is that it makes us reasonable persons, aware of possibilities and capable of solving problems of daily life to a certain extent. Box 6.1 gives various views of people, including scientists, on what science is for.

**Box 6.1** Views on the Purpose of Science

The question on the purpose of science receives various answers from philosophical to practical points of view. Typical answers from those with philosophical points of view are to get a grasp of reality, to ask questions about nature that can be answered by observation, to obtain systematic knowledge, and to explore and discover, while typical answers from those with practical points of view are to obtain knowledge that may be of benefit to humans, to further technology and innovation, and to promote sustainable development. A search on the Internet, for example, turns up a few views as follows:

"The purpose of science is: to validate knowledge in such a way that people with different interests can agree that it is valid" (Adam Nieman).

"The purpose of science is to get a better grasp of reality. It is a process that attempts to produce the best explanations for what we observe" (Todd William).

"To ask testable questions, test those questions, and then revise those questions based on the data" (Carrie Cizauskas).

"Science has three main goals...: To explore and discover, To explain and uncover, To create and make better" (Andre Fahat).

We can gain a further glimpse of views on what science is for from a lecture delivered by Sir John Sulston and John Harris, with participation by Richard Dawkins, in Oxford in 2008. Harris took the view that the basic purpose of science is to do good, while Sulston argued that the purpose is to generate knowledge. They took the subject of synthetic biology as a test case, looking to the day in the future where humankind may be able to guide its own evolution through this new science. Who would decide which way is good for human evolution or even whether it will lead to superior "posthumans"? Can we help nudge evolution in the "right" direction by our own science? With the present inequalities in today's world, would possibilities of enhanced evolution for some lead to even greater inequalities? Both agreed that the question of fairness in distributing the spoils of science is difficult, especially with the present trend in private investment in science and technology, but would look to good social policy and institutions to lead to the right direction.

Similar questions can also be raised about fair distribution of the goods of other areas of science and technology, ranging from health care to energy to information technology. While the basic purposes of science, to do good and to generate knowledge, are generally acceptable, when it comes to real issues in science and society, the devil is in the detail.

Beyond this general answer to the question of what science is for, which applies to everyone, there are deeper answers that apply to people of various occupations, to the scientists themselves, and to society as a whole. Science serves as the main source of knowledge on all kinds of human activities, ranging from basic knowledge on foods, health, and household activities to knowledge on principles of industrial processes and machineries, to knowledge on communication and information. In the previous chapters we examined the role of imagination in scientific discoveries and technological inventions. In reverse, science can be a powerful source of imagination and creativity, not only for further scientific discoveries and technological inventions, but also for paintings, literature, films, the social network, and new models of business. In addition to being a source of knowledge, science serves further as a source of modern technology, which makes use of knowledge it provided to make innovations. Some innovations become subsequently successful through the business acumen of entrepreneurs and the marketability of the commercial products and processes and other factors. In addition to commercialization, science also serves to provide knowledge for various types of occupations, ranging from basic production through industry and services. From the societal perspective, science serves as a tool for various aspects of development, such as agriculture, industry, and services. Development can occur on a short-term basis or can be undermined by unexpected negative consequences. Many paths of development can potentially occur, of which the preferable ones from the broad viewpoint are judged by their sustainability. Science can serve as input for sustainable development paths. From these various functions of science, which we will examine in more detail, we can then give more definitive answers to the question of what science is for.

## 6.2 Science as a Source of Knowledge and of Methods for Problem Solving

An essential function of science is to serve as a source of knowledge on nature, on everything around us, including our environment, and on everything about ourselves in intimate detail. The knowledge contained in science ranges from that about the universe to the subatomic components of matter and, as we saw in Chapter 1, can be divided into various fields such as physics, chemistry, and biology, each of which can be divided further into subfields, for example, organic chemistry as a subfield of chemistry. Some fields have merged with one another to create hybrid subjects, such as molecular biology and nanoscience. Other areas of learning have adopted the methodology and tools of science, such as management science, economic science, and motion picture science. Apart from serving as a major source of knowledge, science also serves as a tool to solve problems, ranging from problems in everyday life to major problems concerning survival and development of humankind. For what purposes is this source of knowledge? It serves a variety of purposes, chief among which are discussed in the following sections.

### 6.2.1 Helping Us Understand the Nature of Things

We are curious and wonder about the nature of things. We explore and gather information on various things, especially those that specially interest us. Science is a source of information that helps us know and understand the world. Information from science comes from research involving experimentation, exploration, and theoretical consideration in detail, covering both the depth and the breadth of the subjects. The results of such detailed investigation are published in scientific journals, after a review of novelty and a check of various criteria by referees, who are peers in the profession. Other results, especially those with technological applications, are registered as some forms of intellectual property. The main conclusions and outlines of the works are simplified and communicated to the public by science communicators and are filtered into classrooms as learning materials for students by educators. The bulk of scientific knowledge with all its details helps us understand various aspects of a nature. It is public record, which has to withstand the test of validity and is subject to improvement by subsequent work. Taken together, this vast record, available as journals, books, articles, audiovisual materials, and even personal accounts, represents the total sum of scientific knowledge. All items stand to be checked for reliability, and cases of bad science need to be rooted out by the scientific community. In addition to intentional deception, a number of reported scientific results are retracted through irreproducibility or errors. Even information that passes the test of validity today may turn out to be inaccurate or wrong in the future and is replaced by new information that represents a better picture of nature. The quest for new and improved knowledge is the major task of scientists. The checks and balances by the scientific community make science honest and valuable for both scientists and the public at large.

Understanding the nature of things is the first step toward various benefits. When we understand the nature of a disease, we can go further toward preventing or curing it. When we understand the structure of matter, we can go on to make new materials with better properties than before. When we understand the reasons why our planet is getting warmer, we can adapt our lifestyles and industries to reduce this warming trend. Application-driven science now occupies the interest of the majority of scientists, following the demands of society. However, the basic purpose of scientific understanding is not just for obtaining these benefits—they are important but nonetheless by-products of the quest for understanding. Curiosity-driven science, sometimes called "blue sky science," is done by a number of scientists who do not have clear applied goals in mind but just want to understand the subject more clearly. Paradoxically, blue-sky research often results in unforeseen benefits, even to greater extents than narrowly applied research. The search for the nature of genetic materials, culminating in the discovery of the double-helical structure of deoxyribonucleic acid (DNA), which opened up so many applications in medical science and biotechnology, is one example of the success of blue-sky research. Another example is the study of the nature of semiconductors, which resulted in the development of electronics and other information-based technologies. We can expect that the current research to find out the basic aspects of neuro- and brain-based science will likely yield unforeseen benefits for human learning and aging and for the development of intelligent robots in the future.

### 6.2.2 Helping Us in Daily Life

We go about our daily routines, mostly without being conscious of how much we are helped by knowledge from science. We choose the food we eat, depending not only on how good it tastes but also on its nutritious value. People who overlook unhygienic foods do so at their own peril. From knowledge that germs can make us sick, we avoid them by keeping ourselves clean and try to prevent ourselves from infectious diseases, such as protecting our skin from mosquito bites. We can look after ourselves in minor illnesses through rest and available drugs or go to doctors, who help bring us back to health again. All these are possible because society makes use of scientific knowledge about health and diseases. We enjoy entertainment and the arts through media made available by science. We use the cell phone, social media, and other products of information science and technology to go about our daily lives. We travel by land, sea, and air, mostly through vehicles developed with the help of science and technology. We do not need to know the principles of how these modern machines work, but it is helpful to be aware of the basic knowledge involved in using them. Scientific knowledge is helpful in taking care of our homes, such as in repair of the electrical, heating, cooling, and plumbing systems and even furniture. Apart from knowledge, science also gives us general methods for solving problems, starting from defining the problems clearly, moving on to forming tentative answers, and going on to test the validity of possible answers (see, for example, Section 1.4, "Turning the Spirit into Action"). We can fix minor troubles ourselves without having to call for outside service all the time. The environment around us, in the city, countryside, and wilderness, needs scientific knowledge and application in order to be sustained and protected from destructive human actions. As members of the public, we can look after the environment, and it helps to know some science, such as how to avoid releasing greenhouse gases, recycle garbage, conserve power, and slow down deforestation. All of these actions and habits come about, directly or indirectly, through the guidance of scientific knowledge. Public awareness of science and technology has become an important issue in many countries, taken up by the government, academia, and civil service organizations. With increasing concern about the state of the environment, safety of consumer products, and other issues concerning science and technology, many individuals and groups have been involved not only in creating awareness of but also in engaging in science and technology so as to ensure public benefit and safety.

However, because science lends credibility to various issues in everyday life, it has unfortunately been cited as an authoritative source of practices and tools that are of dubious value, fraudulent, or harmful. The lure of money, or sometimes just eagerness to lend credibility to doubtful practices or long-held beliefs, makes people cite science as their ally in promoting their causes. It is important to separate the name of science from these practices. Bad science is only a minor portion of all science, which is mostly good in terms of public benefit, but the public and the scientists themselves need to be watchful and root out the problems wherever they are (see Box 6.2).

**Box 6.2** Good Science, Bad Science

Knowledge from science has, in the main, been put to good use. One basic purpose of science is to create awareness in the general public of this knowledge, especially knowledge of relevance to everyday life, so that the public can make good decisions concerning use of scientific information for health, livelihood, or living in general. It is important that mass media, the Internet, and other media be widely available and that the contents, which we might collectively call good science, be correct and helpful. Schools and institutes of higher learning also need to teach good science as general subjects for everyone, not just for scientists and technologists. A good, broad education is needed, which should instill good judgment on scientific information in every student, who will become an educated member of the public. Unfortunately, this has not been achieved in general, and there is a lot of misinformation in the media, which the unprepared public consumes, with bad consequences. Bad science practices are still rampant all over the world, especially in developing countries with poor levels of education.

Many cases of bad science concern health care and nutrition, since the public is vulnerable to issues that are of importance to the state of its health. Ben Goldacre [31] pointed out a number of examples of bad science, ranging from aqua-detox to homoeopathy and pills for "solving complex social problems." These are cases of bad science because they deliberately provide false evidence to try to lend scientific credibility to the practices or products. Some cases, such as in nutrition and alternative therapy, have some real supporting scientific evidence, but this has been distorted or only partly presented.

Not only individuals but also companies and even governments have fallen prey to bad science practices. As an example, many governments, including Mexico, India, Thailand, and African states, bought equipment named GT200 from a company in the United Kingdom, which claimed that it could remotely detect explosives or drugs. Some members of the public and the media investigated this case and found that it operated similarly to a dowsing or divining rod, used since the Middle Ages, supposedly to locate ammunition, gold, water, and burial sites. The electronic sensing system was no more than two sheets of card sandwiched by paper! In Thailand, which was one of the customers, a group of people led by a scientist (Dr. Jessada Denduangboripant) took up the case and called for a public investigation. The government investigated the issue and tested the equipment at the National Science and Technology Development Agency for its ability to detect explosives. The result was that the odds of detecting correctly were equivalent to random chance. The UK government, which at first helped the company sell the equipment around the world, later took up the case in court as fraud. The owner of the company was sentenced to seven years' imprisonment and ordered to pay a fine, which was far less than the value the company obtained from the sale of its fraudulent equipment.

### 6.2.3 Helping Us Solve Present and Future Problems

Our world is full of problems at all levels, from personal to family and work-related problems to those affecting our homes, our towns, and our society. These are related to global problems concerning various issues ranging from hunger to health, from environment to energy. The problems are usually complex, especially those at societal and global levels, requiring various types of action for solutions, including financial, management, legal, and community approaches. Science is often a major component of the solutions to these problems, both because it provides general methods for defining them and forming tentative answers through its approach and because it provides knowledge pertinent to potential solutions. For example, science can provide many tools toward solving problems of the aging society, ranging from medical and psychological help to provision of tools for aid in hearing, seeing, and walking. Traffic problems can be alleviated by regulation through network systems and the provision of best routes through intelligent computer programs that are aided by the Global Positioning System. These are some of the problems the solutions to which need input from science.

The future is uncertain: unexpected events and changes do occur, which lead us down paths that we did not anticipate. However, science allows us to have glimpses of some parts of the future and prepare for things to come. For example, we can expect that the urban population will grow and greatly outnumber people in the rural areas, bringing up big problems such as water and food supplies, energy use, transport, housing, greenhouse gas production, and environmental management. The megatrends can be calculated, and projections can be made, with various outcomes from interventions or forward planning, together with cost projections for these possible interventions. With foresight enabled by input from various sources, including natural and social sciences, we can prepare for the future to come in, say, 10 to 20 years. Beyond that, it will be difficult to make accurate projections. One main factor creating uncertainty of long-term projections is unforeseen changes coming from new science. The energy picture may change dramatically if nuclear fusion becomes a significant source of energy. The future of computing can be affected greatly by quantum computing, which is still undergoing development. Human destiny is shaped by both foreseen and unexpected events, but we can be sure that science plays a big role in shaping this destiny.

## 6.3 Science for Intellect and Wisdom

With science as a source of knowledge and of methods for problem solving, we can see that it serves an ultimate purpose as a quality that promotes our intellect and wisdom in many ways. What we take in through our senses are information and experiences, which we note, learn, and compare with former information and experiences, enabling us to understand the situation at a certain level. Science helps us understand various issues about nature much more thoroughly than our experiences alone because it processes information not only from a single person but also from various sources through systematic investigation over a long period of time. For example, we understand at a certain level that a thunderstorm is a natural phenomenon, a seasonal occurrence. Societies mired in superstition or prescientific beliefs such as those in past history would give different explanations. Some societies believe thunder comes from the wrath of gods. In Greek mythology, lightning and thunder are weapons given to the god Zeus by the one-eyed cyclopes. Many societies in the East have a legend that lightning is due to glitters from a precious glass held by a fairy named Mekla teasing a demon named Ramasura. The sound of thunder comes from the axe thrown by Ramasura in pursuit of both Mekla and the glass. These stories serve as alternative explanations of natural phenomena, valued as parts of cultures of various societies. A scientific explanation of lightning and thunder is based on systematic observation and experiments and forms part of large accounts of physical phenomena that enable us to understand nature in both broad and deep aspects. We can understand that rain comes from clouds, which are composed of water vapor derived from water sources on the earth. Science tells us further that lightning and thunder are phenomena derived from movement of charges in the clouds and the atmosphere. The charges are created by a process that is still not completely understood, generally involving collision between supercooled cloud droplets and small ice crystals moving upward and the larger-size graupel (soft hail) moving downward. The upper parts of the clouds tend to be positively charged and the lower parts negatively charged. The flashes of lightning are traces of the moving charges recombining with one another in the sky or between the sky and the earth, which can be so violent that the air around them expands, and when it collapses it creates the loud sound of thunder.

Explanations of phenomena like lightning and thunder in various civilizations in the past served the purpose in our trying to come to terms with mysterious and awe-inspiring events in nature (Fig. 6.1). They are part of our cultural legacies and have ramifications in ethical and religious teachings. As alternative accounts of natural phenomena, however, the scientific explanation appeals best to our intellect since it helps us understand the phenomena, not as isolated events, but as parts of the whole system of nature. Scientific explanations of lightning and thunder tie in with our understanding of the nature of rain and clouds, of conversion between liquid water, vapor, and ice, and of electricity. These topics came from various fields of scientific studies originally, yet now they come together to give a coherent explanation of lightning and thunder, not as a stand-alone explanation, but as being related to all other issues of related sciences. Lightning can be put in a similar category as battery discharge or lighting in our homes. Furthermore, it gives room for further questions and investigations, such as those concerning the separation and accumulation of charges, the nature of different types of lightning, comparison with volcanic activities, and various radiations other than light emitted from lightning.

**Figure 6.1** Alternative explanations for lightning and thunder. Greek mythology ascribes the events to acts of the god Zeus, the Eastern epic of Ramayana describes them as coming from the demon Ramasura chasing after the fairy Mekla, while modern science explains the phenomenon as a combination of charges separated by storms.

Finally, we can put lightning, thunder, and rain in a broader perspective of their role in hydrology, or the science of water, through association with the water cycle, which involves evaporation from land and ocean water sources, cloud formation in the atmosphere, and rain and snow coming back to earth. The fact that science is accumulated knowledge from various fields of study, woven together to give a coherent picture of the world around us, adds to our intellect. The picture of the world, of the universe, and of ourselves is incomplete—as it is the nature of science to have room for further clarification—but nonetheless satisfying. Furthermore, it is a useful picture in the practical sense. Our knowledge of lightning tells us to be careful when we walk in a thunderstorm and not to wear metallic ornaments that can be conductors of electricity. The airline and building industries can use the knowledge to design airplanes and buildings that can avoid lightning strikes or harm from them. Through science, we also know that lightning is not only a threat to humans and life on earth in general but also an important process that produces atmospheric ozone protecting against harm from ultraviolet radiation of the sun and nitrogen compounds that are used by plants and all life forms. All these realizations come from our intellect that we have gained through science.

We come to the conclusion that in serving as a source of knowledge and of methods for problem solving, an ultimate purpose of science is to increase our intellect, our ability to come to conclusions based on available evidence and reasons. In addition, science also contributes toward our wisdom. As we noted in Chapter 3 (Section 3.2), wisdom is the quality resulting from having a large body of knowledge, experience, and critical awareness, leading to good judgment, prudence, and circumspection. Such quality requires intellect that comes both from the brain and from the heart, that is, it involves both reason and emotion. Science not only promotes our intellect in the narrow sense of giving cold knowledge and solutions to problems but also contributes to the spirit of the heart in the sense that it affects our whole being and attitudes to the outside world. Recall that the spirit of science is born from a sense of curiosity, a sense of wonder, and a quest for exploration. These motives do not just activate our brain but activate our heart as well. The spirit of science is therefore part of the spirit of the heart, and an ultimate purpose of science is to contribute to wisdom as well as intellect.

## 6.4 Science as a Source of Technology and as a Societal Tool

An important function of science, in addition to the quest for knowledge and provision of systematic means of problem solving, is to serve as a source for development of technology and tools for many societal purposes. We noted in Chapter 2 that science as the spirit of this quest for knowledge has given rise to many sparks, some of which are material benefits in terms of devices and machineries and some of which are processes and know-how for making our lives easier and more enjoyable. Science is the main source of modern technology and innovations, and technology, in turn, is the main tool for science, providing the instruments and methodologies for scientific activities. Science and technology are often named as a single entity because of these linkages.

Technology and innovations can have science as a main source in a variety of ways. Science as a quest for knowledge can have technology and innovations as spinoffs through opportunities provided by knowledge. This is an inside-out process, with knowledge generated within science adopted by the outside world through felt needs, including business and societal demands. Of countless examples of such fruits from science, we can just cite a couple. One example of science-generated technology and innovations involves the laser, which is abbreviated from "light amplification by stimulated emission of radiation." A laser is a highly intense beam of light at a single frequency. Lasers are now used in a vast variety of applications, including printers, scanners, optical disk drives, light displays, welding, and surgery. The origin of the laser goes back to 1917, with theoretical work by Albert Einstein, followed by theoretical and experimental works by many eminent scientists, starting from the middle of the past century, including Charles Towne, Arthur Schawlow, Gordon Gould, and Theodore Maiman [27]. The term "laser" was coined by Gould, who noted various possible applications for it, mostly in scientific instruments. Its use subsequently widened to industry, defense, and various service applications with maturation of technologies for its production.

Another example of science-generated technology and subsequent innovations comes from the discovery of a family of carbon atoms linked together to form various shapes [28]. The first such molecule, called *buckminsterfullerene*, or *fullerene* for short (Fig. 6.2), was discovered by Robert Curl, Harold Kroto, and Richard Smalley in 1985. It is composed of 60 carbon atoms arranged into a soccer ball shape, like a geodesic dome designed by the architect Buckminster Fuller. The discovery was an unexpected result of experiments aimed to replicate interstellar matter in the form of long carbon chains. Since then, various other carbon molecules in the fullerene family have been synthesized, a group of which are called carbon nanotubes. These are carbon atoms linked as cylinders of various shapes and many possible layers. Some 10,000 times thinner than human hair, they are stronger than steel, are good conductors of electricity and heat, and have large surface areas. Because of these and other properties, they can be used as materials for various purposes, including touch-screen devices, bulletproof vests, and solar panels and fuel cells.

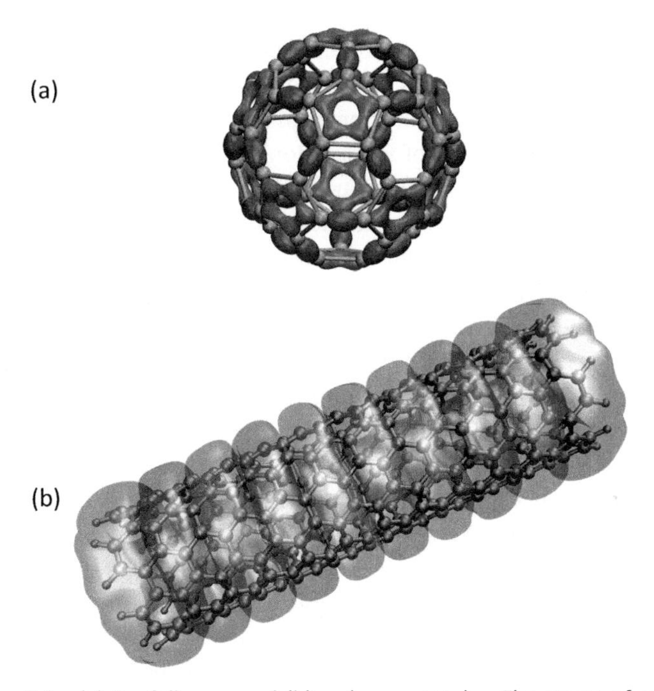


**Figure 6.2** (a) $C_{60}$ fullerene and (b) carbon nanotube. The atoms of carbon are represented by balls joined by bonds, represented by bars and other shaded surfaces. Pictures courtesy of Vidyasirimedhi Institute of Science and Technology (VISTEC), Thailand.

These examples give clear recognition to the importance of scientific discoveries in the generation of technology and innovations, which fulfils social needs and industrial demands. They appear to follow the linear model of innovation, which basically states that innovation results from a linear process, from scientific discoveries to applied research, technology development, and launch of products and processes in the market or to the society at large.

The linear model of innovation is sometimes called the science-push or technology-push model. However, in the past few decades it has been argued that most innovations result from demand pull, namely that it is the demand from industry or society at large that drives the development of products and processes. Closely related to the demand-pull model is the need-based model of innovation, which states that innovations occur as consequences of need felt by society. Examples of mostly demand-pull innovations can be seen everywhere, including in the computer and cell phone industries, alternative energy, and environmental applications such as recycling technologies. Examples of need-based innovations can be found, for example, in vaccines against emerging virulent diseases. Simply put, they are the results of stimulation from outside the sphere of science—namely the society at large—so as to have innovations that are based on demands of the market or needs of society and do not stem mainly just from advances in scientific knowledge in itself. Yet, just like the science- or the technology-push model, the demand-pull model has been criticized for its oversimplification of the complex situation in the real world. Interactive models of innovation have been proposed [29, 30], with interaction between suppliers and users of innovation. Such models should give better representations of science as a source of technology and innovations. In some cases, scientific advances give rise to opportunities that are taken up by industry or the public sector. In other cases demands from the market or needs of society stimulate relevant sectors of the scientific community to focus on the specific problems and come up with new scientific and technological advances.

Both science and technology push and demand pull play their respective roles to different extents according to the characters of the innovations. Many innovations never get past the prototype stage or are shelved even before that stage because of various obstacles (funding, management, scaling-up difficulties, etc.) or lack of mechanisms to pull them forward to the market. Often even when a technology or an innovation is apparently ready for consumers, market uncertainties and public expectation of even better technologies around the corner can obstruct its path of development. On the other hand, even with clear societal demands, some needed technology or innovation cannot materialize because the relevant science is not yet ready. Energy production from nuclear fusion technology is one example; effective therapy for many types of cancer is another example of such needed technologies. In conclusion, for society to reap the benefits of innovations, a number of factors, which can be grouped together as science/technology push and demand/need pull, have to come into play and have to interact well at the opportune moments. In being the source of knowledge in general, science serves society also in bringing benefits from such knowledge, through technology and innovations in which it plays a big role, but realization of the benefits requires many other interacting factors.

## 6.5 Science for Benefits

We can come to the conclusion that in addition to building intellect and wisdom, an important purpose of science is to obtain benefits from it. In some cases, the benefits were obtained as a by-product of basic science, or studies intended to gain knowledge and understanding of natural phenomena. In other cases, the benefits are the primary aims of the studies, which are therefore collectively called applied science. Technology and innovations often result from applied science, although they involve many more parties other than scientists to bring them about. Successful technology and innovations likely require interaction among many players, ranging from scientists to entrepreneurs, industrialists, business managers, and finance and legal professionals. Figure 6.3 depicts a simplified diagram showing science as the innermost core of an onion-like structure, with the outer layers as technology, innovation, competitiveness, and a healthy economy. Each layer requires players other than science in order to function. Technology and innovation require engineering capability, business acumen, entrepreneurship, good finance, and legal capability among other factors. They operate at the individual or firm level, while competitiveness operates at both firm and country levels. With good public policy, science, technology, and innovation can also be used as a tool to achieve inclusive development of society, which will help achieve sustainability as an outcome. In addition to good policy, good governance, an educated workforce, market efficiency, and a good macroeconomic environment all help promote beneficial outcomes for science and technology.

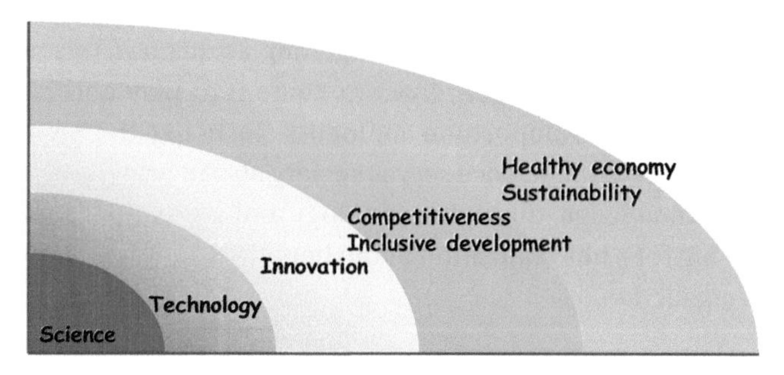


**Figure 6.3** A simplified onion-like structure showing science as the core, contributing to technology, innovation, competitiveness, and a healthy economy. Depending on the purpose of utilization, science, technology, and innovation can also lead to inclusive development, in addition to competitiveness, and to sustainability, in addition to a healthy economy as an outcome. Factors other than science and technology, such as entrepreneurship, finance and legal infrastructure, and consumer acceptance, are essential for the outer layers to function.

Of countless examples, we can trace the development of the Internet as a prime case of the development of science and technology intended for benefits. Transmission of messages as text, voice, or pictures became possible over the past two centuries with the development of the telegraph, radio, and television, although still as one-way communication or limited two-way communication and not an instantly interactive one. Development of electronic computers in the 1950s and their communication through computer networks paved the way for the development of the Internet, which makes instant interactive communication possible between individuals or groups all over the world. It was first initiated in the 1960s as a project for communication among a small number of members of a network called ARPANET (ARPA is short for Advanced Research Projects Agency, an agency of the US Department of Defense). The key idea that makes this instant, interactive communication possible is to transmit messages in small packets through all the members of the network through designated gateways so as to arrive at and be reconstructed into whole messages by the intended receiver. This brilliant idea called packet switching, proposed by Paul Baran and

Donald Davies, ensures that error-free, instantaneous messages can be transmitted among network members. Baran and Davies also suggested the concept of distributed networks with no centralized nodes, a feature that makes the Internet a truly remarkable invention in line with the ideas of democracy. The first e-mail program was conceived by Ray Tomlinson on the ARPANET. The network was expanded with support from the US National Science Foundation, resulting in NSFNET with a standard working protocol in the 1980s. Commercial Internet service providers emerged in the late 1980s, marking the passage of the technology users from government to the private sector and the public at large. Development of the World Wide Web by Tim Berners-Lee, depending on the use of what is called hypertext, enhanced the usage of the Internet even further. The technology caught the interest of society the world over, spawning the development of online business, online learning, and social networks like Facebook and Twitter.

The Internet ranks as one of the most important human achievements made possible by science. Although its core is electronics, information science, and computer science, it is made up of many other contributing sciences, including physics, complexity science (study of complex systems), and life sciences (neural networks, behavior, and evolution). It is now a main tool of all disciplines, including social sciences and humanities. Its use in business, government, entertainment, sport, and virtually all spheres of human activities has remarkably changed the nature of work and lifestyles. The benefits to society of this single technology bring to mind the landmark report by Vannevar Bush, *Science, the Endless Frontier*, published in 1945, urging the US government to support science, including basic research (see Box 6.3). Significantly, Bush was also an early pioneer of what eventually became Internet technology, demonstrating from his own work that the benefits and knowledge obtained from science are not far from one another.

**Box 6.3** The Endless Frontier Revisited

Do the benefits of science stem from intentionally targeted R&D, or are they the by-products of our quest for scientific knowledge? This long-standing question has immense implications for science policy and for economic and social development in general.

In 1945, as the Second World War was ending, Vannevar Bush, director of the US Office of Scientific Research and Development, submitted a report to President Franklin Roosevelt, entitled *Science, the Endless Frontier*. This was a report in response to the president's request for recommendations on the diffusion of scientific knowledge gained from the war effort, the direction of science in the fight against diseases, and the support for R&D of scientific talents in general. The report called for broad-based support for science and the creation of a single federal agency, the National Science Foundation, to fund basic research in all areas, ranging from medicine to weapon systems. Partly because of successful war efforts, which saw the development of the atomic bomb, the radar, and penicillin, all starting with basic research, the recommendations were well received. Although the created agency turned out to be far smaller than what Bush envisioned, his report started a new era of generous support to science from the government, which lasted for around two decades. Science, including basic science, was designated as the endless frontier, the discovery of which should usher in solutions to various problems of society. In recent years, however, the concept of benefits to society of basic research has been scrutinized, although its importance has been retained in most advanced countries. New terms such as "transformative research" have been proposed. A European vision for research and innovations for the 21st century is called "Society, the Endless Frontier," reflecting the emphasis toward the needs of society.

Bush's outlook on the use of science was reflected in his 1945 essay entitled "As We May Think," in which he proposed the concept of the "memex" machine, which would act as collective memory of all knowledge. He based his proposal on the technology of high-resolution microfilm, considered to be advanced then. The advent of the more powerful Internet technology, which the National Science Foundation helped launch, has made his proposal close to realization.

We can make an analogy of science and technology with a fruit tree (Fig. 6.4). Basic science can be compared to the roots, stem, and branches, which are inedible but essential for the tree to survive and grow. Applied science can be compared to the leaves and the flowers, which will eventually be pollinated. Pollination requires outside factors such as wind, insects, or other animals, just as a scientific and technological undertaking requires contribution from external agents in order to achieve success. Technology can be compared to the resulting fruits that are picked and eaten. Although we cannot eat other parts of the tree, we need to tend to them all as otherwise the fruits cannot be produced.

**Figure 6.4** Science and technology as a fruit tree. Basic science can be compared to the roots, stem, and branches; applied science to leaves and flowers to be pollinated; and technology to fruits that can be picked and eaten.

Many people think that developing countries can take shortcuts to economic and social development through fostering of technology alone without having to support basic science. In this way they can take a shortcut to development. The argument is that basic science is already done in developed countries, and doing it also in developing countries would be repetitious and a case of doing too little, too late. This line of reasoning has some truth, since it makes no sense to repeat what other, more advanced countries are doing, especially with the little resources available in developing countries. Even when the research is original, it will likely not be competitive because of lack of personnel and resources. It is better—so the argument goes—to have a catch-up strategy, bypassing the basic research and going straight to the application phase. Other people go even further, saying that the developing countries need only to import technology from abroad without having to invest in it themselves. The strategy is analogous to buying fruits from the market without planting the trees for homegrown fruits. However, this strategy and even the catch-up strategy that have merits in allowing shortcuts to the development of science and technology in developing countries have serious drawbacks if applied as a blanket approach. They may work in some cases of industry but even then only on a short-term basis. Once the economy grows to a certain stage, some home-grown technologies will be needed, both for a measure of self-reliance and to contain the cost of technology purchase. Furthermore, many problems requiring science and technology solutions, such as tropical diseases and tropical agriculture, may be unique to the developing countries, without solutions to be obtained from overseas. There is therefore a comparative advantage of doing some basic and applied research in the developing countries themselves, in addition to just importing or catching up with available knowledge and technology from outside. Home-grown technologies, which may be adapted from imported ones or developed entirely from indigenous sources, are also more appropriate, considering the infrastructure, raw materials, and local markets. In short, we should plant some fruit trees, not just rely on buying fruits from the market.

The benefits of science fall under many levels, ranging from persons to groups, societies, and the whole world at large. At the aggregate level, we can talk about the benefits of science to humanity and to the environment. As noted earlier, we should also be aware of potential threats that science may pose to humanity and the environment as well, both from intentional and from unintentional actions. In our discussion of the purpose of science, therefore, we should consider its contributions and implications to humanity and the environment.

## 6.6 Science for Humanity and the Environment

Science is meaningful, not only in the sense that it contributes to the stock of knowledge and yields various benefits, but also for its role as common property of humankind. As such, it has accumulated through human history, especially over the past few centuries, similarly to arts, literature, architecture, and other creative works collectively called culture. These creative works, in which science features prominently, contribute to humanity and distinguish us from other animals. The creative works of science are different from others in the sense that they are not solely made by any one person or group but are a common heritage based on accepted principles. These works can extend past achievements or sometimes overthrow and replace them with new and improved principles or gadgets. New findings in science are always related to existing principles and contents, helping give a clearer picture of nature and better technologies for our benefits. In contrast, other creative works, such as paintings, novels, or plays, are all individual works, and although they can have effects on subsequent works by others, they can only do so through personal influence. Even though a painting may be a cooperative creation of a group, or the work of students of a master, it is still a single piece of work separated from others. On the other hand, a scientific work, after its accomplishment by the creators, becomes the common property of humanity. Creation of new scientific work is not a free creation independent from others but an extension or improvement of existing work done previously by others so as to build the knowledge as a common human heritage further. As Isaac Newton said, "If I have seen further it is by standing on the shoulders of giants." A comparison might be made of science to an anthill, where every ant contributes through the addition of small soil granules so as to help make it into a large hill as a common property of all the ants.

A large part of science is the common property of humankind, from which everyone stands to gain benefits. However, new knowledge gained from research and development (R&D) is intellectual property that may be patented or protected from free use by legal means. This is so that the researchers can gain benefits from their work and industry can make and market the products without fear of being copied. Protection of intellectual property is more common for the results of applied science and technology than for those of basic science. Even then, such protection will be valid over a certain period, after which the invention or discovery becomes common knowledge accessible to all.

We have so far looked mainly at the positive side of science, science for humanity. Unfortunately, there is also the negative side, science without or even against humanity. Mohandas Karamchand Gandhi had a list of seven social sins, sometimes called the seven blunders of the world, of which science without humanity is one. There are many examples of events in the past where science was used against or without humanity. Science has been used to build instruments of war, aggression, torture, and other antihuman purposes. Some people tried to use genetic sciences to build superhuman races for their own advantage. Others use scientific tools to cheat and perform illegal acts such as money laundering. In some cases, such as weapons, arguments can be made for using them as deterrents against those with ill intentions. Research into hacking of computer systems is said to be necessary in order to prevent such acts by ill-intentioned people. We will discuss these complex issues of the dark side of science in Chapter 9.

In addition to promoting humanity, science also has a big role in conserving our environment and helping us reach the goals of sustainable development. As noted earlier (Chapter 2, Sections 2.5 and 2.6), there are planetary boundaries beyond which humanity should not exceed, as otherwise the environment could be disturbed toward the state of unsustainability. The concerns include climate change, ocean acidification, stratospheric ozone depletion, freshwater use, land use, and chemical pollution. In many ways, science can be blamed for creating part of the problem, since these threats to the environment occur partly as consequences of changes in human activities caused by scientific development. However, solutions to problems of the environment can be found in science. New areas that would help solve the problems include sustainability science, an interdisciplinary science aimed toward understanding of human–environment interactions, and design and implementation of strategies that would promote sustainability of the environment.

# Chapter 7

# Present and New Challenges

*Who among us would not be happy to lift the veil behind which is hidden the future; to gaze at the coming developments of our science and at the secrets of its development in the centuries to come?*

—David Hilbert, *"The Problems of Mathematics,"* presented during the course of the Second International Congress of Mathematicians, Paris (1900)

Science has contributed to economic and social benefits through discoveries and innovations that led to developments in trade, industry, agriculture, communication, education, health, and the quality of life in general. It is expected that we will make increasing advances in solving challenges from within science itself for finding new knowledge and deriving new benefits. However, many serious societal challenges remain to be solved with the help of science. Despite advances in medical science and agriculture, hunger and diseases still remain important threats, especially in the poor and war-torn regions of the world. Good education is still out of reach for most people. Challenges remain in providing sufficient services for children, women, and disadvantaged people. Furthermore, the changes in society, partly brought about by science, bring about new challenges, such as a change in the nature of jobs, an increase in the number of the elderly requiring assistance, and threats from cybercrime, cyberterrorism, and other misuses of

*Sparks from the Spirit: From Science to Innovation, Development, and Sustainability*
Yongyuth Yuthavong

ISBN 978-981-4774-57-4 (Hardcover), 978-1-315-14599-0 (eBook)
www.panstanford.com

information technology. Science needs to be responsive to these new challenges. Furthermore, sustainability should be sought in the outcome of innovation and development. This would require an adjustment of our ways of life toward moderation in consumption and consideration of long-term effects of use of the fruits of science and technology. Sustainability science, a new science concerning sustainable development and lifestyles, can guide us in making such adjustments for transition to a stable world.

## 7.1 A Scorecard for Science

Looking back, we see that science has done well in contributing to our stock of knowledge and in helping create benefits for society. In contributing to our stock of knowledge, science emerged many centuries ago as a subject separate from philosophy and the arts of the ancient Greeks and freed itself from astrology, alchemy, and other subjects, with its own distinction of using reason and systematic approaches to investigate natural phenomena. Present-day science gained from science and technology of many civilizations, including those of ancient Greeks and Romans, like a river arising from various tributaries. The view of our place in the universe has been altered from the knowledge that it has accumulated. Rather than being at the center of the universe, both where we live and who we are, we now know from science that we are only a tiny part of it. Our planet is only one among many orbiting the sun, which is a star among billions of stars in our galaxy, which, in turn, is just one among some 200 billion galaxies. As a living species, we may be unique in possessing high brain power and other abilities that make us dominant today, but like other species we are a product of evolution with its rules, which all living species, including us, have to obey. If we are not prudent about the way we live on this planet, together with other living species, and about the environment, we could fall by the wayside and become just another extinct species. At the level of separate subjects—physics, chemistry, biology, and others—science with mathematics as its key ally has achieved tremendous progress over the past few centuries, as highlighted in Chapter 1. If we are to score science for its achievements in gaining knowledge for humankind, we would surely give it an A.

Judging from the role in helping give rise to technology and innovations, we should also give science an A for the subject "science for benefits." Benefits in health can be witnessed from our ability to deal with diseases and promote a healthy lifestyle with the aid of medical science. Life expectancy has more than doubled since 1900 to around 70 today. The fruits of science can be compared to the happy ends of fairy tales or the stuff of science fictions. To many, science is like magic, even better than fairy tales. We do not need a magic carpet that can accommodate only one or a few, but we fly in airplanes that can carry 500. Unlike Hansel and Gretel, we can use the Global Positioning System (GPS) to find our way home and to go places. We can see and hear from afar, either in real time or from recordings. Thanks to modern agriculture and food industries, many people are well fed and have a choice of good foods from anywhere in the world. Homes in many countries are built with materials and supplied with utilities that give comfort and convenience. We can connect with people and places around the world, chatting, sending messages, and enjoying entertainment from the Internet and other technological means. True, science did not accomplish all these by itself but needed the help of other allies, including engineering, business, industry, and management. Without science, however, we could not have achieved all of these feats.

Things take a different turn, however, when we look at another area of concern of science, namely science for development and sustainability. We will concentrate on world social, environmental, and economic development, not just on technological development stemming from science, as already discussed in Chapter 2. In addition to giving benefits to those who can afford them, science as a human heritage has an obligation to improve the lives of everyone in the world. Up to now, people in less developed countries, which comprise the majority of the world population, have not benefited from science to the extent to which they should be entitled. The people at the base of the pyramid, as the low-income people are grouped, both in developing and in developed countries, still live in hunger, and the food that they eat is not as nutritious as it should be, in spite of the fact that the world as a whole has a high capacity for food production. They live in dilapidated areas and are denied the basic necessities of their homes. They suffer from poor living standards and inadequate health care and education.

Women, children, the elderly, and disadvantaged people are not given opportunities to meet their special needs. The benefits of science and technology, with all the wonderful gadgets and services, are denied to the majority of people in less developed economies because they cannot afford them. It was noted [32] that technology lacked a "human face" and did not help alleviate problems of poverty and unemployment. Science as a key ally of technology must also share the guilt and must be given a score of F, for failure to be of substantial benefit to the majority of the people of the world who still live in poverty, with poor health care, and in crumbling cities and communities. It must be given an F also for not yet being able to stop the planet from sliding into an unsustainable state with regard to the environment. Indeed, many blame science as a culprit, together with technology and industries, in causing unfavorable climate change, pollution, and other problems of the environment. Together with other allies, including environmentalists, inventors, and politicians, scientists must try to improve the state of the world (Fig. 7.1).

**Figure 7.1** "Am I going to die, doctors?"

However, the past few decades have seen some changes in the right direction. Cell phone technology and the Internet, for example, have helped connect people all over the world. Technology, as a whole, has played an important role for some developing countries to emerge from the poor income status, and some countries have even entered the high income status, thanks to development strategies that put science and technology at the forefront. Globalization, made possible by science and technology advancement, has helped many developing economies, for example, by distribution of jobs and businesses. Many developing countries also adopt the strategy of inclusive growth, with the aim of uplifting people at the bottom of society. Yet the accusation that science and technology have not been mobilized enough for global development still stands, and those involved must do more in order to change the grade from F to a passing one. This subject will be discussed further in this chapter and Chapter 8.

We come now to the subject of risks from science and their mitigation. Science has been used many times in the past, intentionally and unintentionally, in ways that pose harm to people or the environment or in ethically questionable ways. To the public, the science-gone-wrong image of Frankenstein comes to mind. We can distinguish many types of risks from science, including risks of physical harm, risks to health, risks to the environment, social risks, and ethical risks. The discoveries of nuclear physics have opened up a potential supply of useful energy but have also led to threats of nuclear weapons and accidents of nuclear power plants. Drugs give us cures from diseases, but they can also have adverse, even life-threatening, effects. Pesticides and chemical fertilizers help boost agricultural productivity but pose a threat to the environment from indiscriminate use. Use of insecticides like DDT can help us avoid insect-borne diseases, but they become a threat to birds and other wildlife species. Social risks range from use of technology in criminal activities to addiction to designer drugs. Ethical risks are posed when science makes it possible to do something new that has implications on moral judgment, for example, creating new life forms, cloning humans, or transplanting stem cells. The risks from science are sometimes foreseeable, as in making nuclear power plants that may encounter catastrophic accidents in their operation. However, since science is engaged in finding new knowledge, the associated risks may not be foreseen. It was not foreseen that insecticides could have an effect on bird reproduction, nor that the use of thalidomide as a sedative by expectant mothers would result in severely disabled babies. Lessons have been learned over the years, and there are now preventive and mitigation plans for the potential risks from various science-based activities, especially those that concern humans, living species, and the environment in general. Science needs good governance—prudent decision making, including risk assessment and mitigation—both by the scientists themselves and by society, which is impacted by its effect. Measures of risk control and mitigation have become much more stringent than in the past, such as in experiments on human subjects that now need rigorous procedure and oversight. This subject is still evolving, and rather than giving a final grade for risk assessment and mitigation in science to the people involved, we should give it an I (incomplete) or an IP (in process) for the time being. We will return to this subject in fuller detail in Chapter 9.

## 7.2 Challenges to Science

We have seen that science has been doing well in gaining new knowledge and producing benefits to society. However, the benefits have tended to accrue to those who can afford them, especially those in the developed countries, since they mainly come in the form of commercial products and services and in the form of public goods that require significant investments. The challenges to scientists and those concerned with development are therefore to address the inequalities in access to the benefits that science generates. These may be addressed through ensuring fair benefits from the products and services generated from science. The international communities can make special considerations for poor countries or communities, especially where the benefits concern vital issues such as health and hunger. In the cases where the products and services that result from science, such as vaccines and drugs, cannot be afforded through normal market mechanisms, special provisions should be made so that access to them is not blocked by an inability to buy. A Facebook post in 2009 saying, "No one should die because they cannot afford health care and no one should go broke because they get sick," or related versions, has been shared by more than a million people, showing that the social network overwhelmingly shares this opinion. The broad picture of the challenges for science is to make it a vital and affordable tool for sustainable development, aiming to end hunger and poverty, to promote health and equality, to build cities and communities, and to make the human society and the environment sustainable in the long run. Special efforts should be made so that people in developing countries can have greater shares in the benefits that science generates. Scientists all over the world, including those in the developing countries, should try to make sure that where their work can generate benefits for humankind, those benefits should be accessible to the poor and the disadvantaged.

Challenges also come from within science itself, concerning what to pursue and how to proceed in the future. Many areas of science carry "lights and shadow," as has been noted by Koji Omi, the founder of the Science and Technology in Society Forum held annually in Kyoto for scientists and world leaders to discuss the issues of benefits and risks of science and technology for humankind. Another broad challenge for science is what proportion of effort should be made in basic science versus applied science, technology, and development. Yet another major challenge is to address the imbalance in science capacity between the developed and the developing world, resulting in a situation where the developing world, which carries the majority of the human population and the burden of development, still makes a tiny contribution to science to solve its problems.

## 7.3 Challenges from within Science

### 7.3.1 Is Science Finished?

The past century saw breathtaking advances in science, ranging from basic knowledge of nature to unprecedented applications to human society. Fundamental laws of nature were revealed by Albert Einstein's theories of relativity, which replaced the classical laws of motion proposed by Isaac Newton two centuries earlier. Exploration of the universe, informed by these theories and with the help of new instrumentation, led us to realize the nature of the expanding universe. At the small scale, the nature of matter was revealed by quantum mechanics, which informs us that it has the quality of both waves and particles at the same time. Atoms were found to consist of protons, neutrons, electrons, and other particles, all of which, in turn, were made up of more fundamental particles. Mass and energy were shown to be interchangeable, and a study of radioactive decay led to the discovery of nuclear energy. A study of electronics led to the development of computers and their various applications. The Internet and other applications of electronics and communication technologies led us to the present age of smart devices, including robots. The science of aviation led us to the age of air travel and anticipated travels in space. Development of chemistry led to many types of useful materials, ranging from drugs to plastics. Discovery of antibiotics from microbes, and subsequently also from chemical synthesis, led to major breakthroughs in treatment and prevention of diseases. Vaccines were developed against many important diseases, and after eradication of smallpox we are now on the verge of eradicating polio and other major diseases. Discovery of the structure of deoxyribonucleic acid (DNA) led to the age of molecular biology, with new insights into the nature of life and the ability to produce new drugs called biopharmaceuticals. The word "DNA" is now widely used, not only in science, but also in all areas of human knowledge, to mean the fundamental element of things.

Altogether, science has achieved so much during the past few decades that some people ask whether there is a limit to knowledge that science can create. Indeed, the end of science [33] was anticipated, in the sense that major new discoveries like evolution, relativity, and DNA structure are no longer forthcoming. While other smaller discoveries will still be made under the umbrellas of previous major ones, it was argued that science has reached its domain limit. New fundamental proposals such as the anticipated theory of everything, even if realized, would be speculative, philosophical, and impossible to prove experimentally. While some people may be swayed by this argument, it should be pointed out that major discoveries, or what we might call "disruptive" discoveries, are by their nature rare. As history shows, truly major discoveries, which change the paradigms and start off new fields of science, may come only once or twice in a lifetime. The past century has seen a few major discoveries that truly transformed the basic principles of science on which modern science is now based. Records show that far from dying down, scientific discoveries continue to gather pace, as witnessed from the number of publications and their impact on technology and innovation. Pessimists might say that such increase might represent the flurry of activities that are like flare-ups of dying stars before they reach the end of their existence. On the contrary, judging from the nature of science, in that it thrives on experimental evidence that helps build new theories, the gathering pace of scientific publications can be seen as evidence of an increasing stock of knowledge that will contribute to new major breakthroughs to come in the future, like gathering storm clouds that will inevitably result in major downpours. New advances in instrumentation and computation will further help in increasing the momentum of advances in experimental science, and artificial intelligence will help in meeting new challenges in solving complicated problems requiring data processing of complexities beyond human capability. Issues discussed by leading scientists and those concerned with the applications of science anticipate lively development in various areas of science, ranging from cosmology to cognitive science and molecular biology [34, 35]. Far from being finished or nearly finished, science is thriving as never before.

### 7.3.2 What Kinds of Science?

Science is a vast subject. It can be subdivided according to specific disciplines, such as biology, physics, and chemistry. In the past few decades, the boundaries between various specific areas of science and technology have tended to be blurred, and there are many hybrid areas of science resulting from the convergence of these areas, such as bioinformatics, nanoscience, and molecular biology. Currently, four main streams of science are growing rapidly, with the tendency to be merged with one another. They are characterized by the prefixes nano-, bio-, info-, and cogno-, often shortened as NBIC [36]. "Nano-" concerns small-dimension aspects (around a nanometer or one billionth of a meter) of science, "bio-" signifies the study of life, "info-" emphasizes information and computation aspects, and "cogno-" deals with the cognitive aspects such as the nature of intelligence. Examples abound of convergent sciences and products derived from them. Robotics is a field of study that sees the convergence of mechanical engineering with computational and cognitive sciences. Genomics combines molecular biology of the genes with advanced computational capabilities in order to map the structure and elucidate the functions of the genes and associated DNA. As a typical example of products arising from the convergence of various areas of knowledge and technology, we can consider the electronic nose. This is a device that incorporates features derived from chemical analysis of volatile compounds, delivering samples for detection through a bioreceptor system similar to the biological olfactory system and a suitable electronic sensing mechanism to identify the compounds and their amounts. Many other products from convergence of various areas of knowledge and technology include the electric car (mechanical and electrical engineering, materials science, sensing technology), transplantable organs (medical science and engineering, immunology, printing technology), and intelligent homes (sensing technology, mechanical and materials engineering, the "Internet of Things").

It should be noted that when science started as an area of knowledge, back in the days of the ancient Greeks, there was no clear distinction between various branches of science. The branching of science into various distinct areas only started much later, when the contents of each area grew so large. The recent blending of various areas of science only shows the unity of the principles of science. The current convergence of science and technology fields is seen to lead to the evolution of convergence at a higher level, namely convergence of knowledge, technology, and society [36].

Science can also be divided along the line of the aims that it intends to achieve. Hence basic science is aimed at gathering new knowledge, and applied science is aimed at using the knowledge for benefits to human society or to the environment. Technology, a subject on its own, is linked to science, especially applied science in that it provides know-how or tools to achieve such benefits. "Development" is a general term denoting healthy growth, such as that of society. As discussed earlier, large parts of technology and development are sparks from the spirit of science. One of the main challenges to science is the proportion to which efforts should be made toward these various facets of science and its offshoots. As discussed in Chapter 6, a model for science policy is to support basic science on the past experience that it led to applications, such as seen in the development of communications technology, drugs, and various products. The trends in the past few decades, however, are for more application-driven research and development (R&D), while basic science is under pressure to demonstrate its cost-effectiveness, especially in highly specialized theoretical subjects. A balance needs to be made on the distribution of efforts, in terms of both people and financial resources, to pursuits in basic science versus more applied areas. It should be noted that because of its nature, there will always be uncertainty as to whether or when basic science will bear fruits for society. Experience shows that when it occurs, applications of basic science tend to be broad ranging and can have a huge impact. Modern medicine, for example, is making increasing use of basic genomic science in diagnostics, prevention, and therapy. Advances in solar technology and electric vehicles also stem, in large part, from basic physics and electronics. Developed countries invest a significant share of their budgets in basic science, although larger shares are allocated to applied science, technology, and development, which require larger resources by their very nature. Even developing countries, with scant resources, should devote a part of their efforts toward basic science. This is a prudent policy, since it will strengthen education as well as provide a broad window to new scientific development in the world. Some developing countries are in the catch-up phase of development, and new basic knowledge will help them in devising good strategies for technology catch-up.

### 7.3.3 Who Should Share in Doing Science?

Science is a universal subject. It explores and gathers knowledge, not for any group of persons or countries in particular, but for all of humankind. With the knowledge gathered, we can go on to build technology and innovations and further to develop economies and societies. It is important therefore that science be available to the global community and the global community contribute to its advances. However, unequal development in various parts of the world has resulted in lack of access of the majority of the global population to science and its benefits. People in developing countries, the majority of the human population, do not have the education or material resources needed to be able to access scientific knowledge that would help them in their livelihood or daily lives. Importantly, the majority of scientists and technologists in developing countries cannot participate meaningfully in advancing scientific knowledge, both for the benefit of humankind in general and to help alleviate the problems of their own countries in particular. Typically, developed countries spend 2%–4% of their gross domestic product (GDP) for R&D activities. Developing countries, due partly to other pressing needs and to poor scientific infrastructure (few scientists, lack of equipment, etc.), typically only spend less than 0.5% of their GDP for R&D [37]. Given that the GDP of these countries is already small, these data serve to show that the developing countries are generally not investing in the generation of knowledge and therefore are in danger of falling even further behind. Although countries can gain knowledge through transfer of technology and upgrade of education systems, R&D expenditure is a measure of how much the countries are proactive in mobilizing science and technology. We can therefore conclude that developing countries still have little share in the generation of scientific and technological knowledge. This situation should be corrected for sustainable development to be achieved by the developing countries, since this requires their increased capacity in solving their own problems, as well as in helping solve global problems such as poverty, hunger, and health, which are to a large extent the problems of developing countries.

There are international programs that boost the contribution of scientists from developing countries, such as the International Foundation for Science registered in Sweden, the International Development Research Centre created by the Canadian government, and the South-South Collaboration Programs of UNESCO and other UN agencies. Supported by UN member states and host countries, the International Centre for Genetic Engineering and Biotechnology, with main sites in New Delhi, India; Trieste, Italy; and Cape Town, South Africa, is dedicated to advanced research and training in molecular biology. The Abdus Salam International Centre for Theoretical Physics, also in Trieste, has been the main center devoted to advances in theoretical physics with contribution from scientists from developing countries over many decades. Universities such as the Okinawa Institute of Science and Technology and the King Abdullah University of Science and Technology offer research and education programs with scholarships for international participants. These (Box 7.1) and other programs represent a few examples of international efforts to enhance the participation of scientists from developing countries. Global collaboration of scientists, including those from developing countries collectively called the South, in international research aimed at problems of health, agriculture, and other problems of the South will be discussed further in Chapter 8. More efforts in this direction should help enhance the role of developing countries to help themselves and the global community through scientific advances.

**Box 7.1** The Developing World Sharing in Scientific Endeavor

The majority of the world population lives in developing countries. Science helps solve challenges of development, and it should be expected that scientists of the developing world have a major share in solving problems such as tropical infectious diseases, hunger, low agricultural output, and lack of innovation for the poor. Yet, because of the lack of infrastructure for support of science, including the lack of laboratories, equipment, funding, and funding support, little scientific research is done in the developing world. Scientists from the developing world, many of whom are trained in universities and institutes of developed countries, have little choice but to stay and work there because of lack of opportunities back home. This phenomenon of "brain drain" exacerbates the problem of low scientific participation in developing countries. Because of the universal nature of scientific endeavor, universities and research institutes in the developed world normally have an international character. However, many have no or little linkage with the home countries of scientists from developing countries to enable them to eventually build their careers in their country of origin. Through the effort of international organizations and those in some developed countries, international research centers and programs have been established with a significant participation of scientists from the developing world. These centers and programs typically host the scientists who normally work in the developing countries, or will eventually go back there, in collaboration with those from the developed countries so that linkages are formed in addition to joint scientific output.

The Abdus Salam International Center for Theoretical Physics in Trieste was set up with sponsorship from the Italian government, UNESCO, and the International Atomic Energy Agency (IAEA), through the initiative of the late Nobel Prize winner Abdus Salam. It hosts physicists from the developing world who can periodically come and work with leading world physicists on various problems in physics in a creative surrounding. Research areas include high-energy physics, cosmology, mathematics, earth systems, applied physics, and quantitative life sciences.

The International Center for Genetic Engineering and Biotechnology has three main sites: Trieste in Italy, New Delhi in India, and Cape Town in South Africa. It was established as a research and training center with promotion from the UN Industrial Development Organization (UNIDO) to find solutions to problems such as prevention and therapy of tropical diseases, plant biotechnology, and biofuels. Scientists from developing countries are supported to work there or in collaborating laboratories in the developing countries themselves. Interestingly, Thailand was first chosen by the selection committee to be the site of the international center before the decision was reversed by a strong campaign by India, prompting Thailand to set up its own national center, which has played a large role in developing the technology in the country.

The Okinawa Institute of Science and Technology is an international school supported by the Japanese government, offering PhD programs in areas such as neuroscience, cell biology, ecology, computational science, physics, and chemistry.

King Abdullah University of Science and Technology located in Saudi Arabia offers research and graduate education programs in biological and environmental science and engineering, physical science and engineering, mathematics, and computer science.

## 7.4 Challenges to Science from Societal and Environmental Problems

Knowledge generated by science helps bring benefits to humankind through technology and innovations. Many benefits are made in terms of products and services that bring in monetary returns to the people involved. However, science helps not only in generating financial profits but also in generating social returns. Indeed, in many cases the economic returns are just part of the bigger picture of the social purposes.

Drugs, vaccines, and other medical products and services are for the health and welfare of people. Commercial benefits from medical science are not its main purpose but provide a mechanism for the products and services to reach the people who need them.

Governments can play an important role to make sure that the public can afford health care through measures such as provision of basic health welfare and a policy for health insurance to cover every member of the public. Nongovernment agencies can also play important roles in helping the public obtain health care. Various other mechanisms can come into play in generating and using the benefits from medical science, including import and export of medical products and services. Science is the generator of knowledge, which is then transformed by the market mechanism, the public sector, and other concerned actors into products and services as parts of the health care system for the benefits of the people. Similar things can be said about the production of food. Thanks to agricultural and related sciences, the capacity of food production is potentially sufficient for the world population, although many people still go hungry because of poverty, lack of access, poor government policy, or a combination of these and other factors. A large part of agricultural yield has also been diverted to energy production. Both the government and the private sector have roles to play in R&D on food and other agricultural products, with the former putting more emphasis on commodity products for the masses and the latter on products that bring in profits. Farmers also do their own experiments in improving the yields and varieties of their products. The main conclusion is, again, that science has a big role to play in bringing food to people. However, its role is mainly as a generator of knowledge, which must then be taken up and transformed into public goods by many other actors.

The tools of science can be mustered to help solve problems in health and food and to meet other societal needs, including education, housing, social welfare, and various public services. Some new tools are generic in nature and can be applied to health, agricultural, and other problems. The science of genetic modification, for example, has advanced to the stage of allowing editing of the genome of various organisms, including humans, so that in the case of medical application, it will soon be possible to edit abnormal genes causing diseases and hence cure the patients through what is called gene therapy. Genome editing can also be applied to problems such as elimination of insect pests and improvement of plant varieties. Advanced research is ongoing for prevention and therapy of diseases such as dementia, heart disease, and diabetes through using appropriate stem cells. Box 7.2 gives further information on the status of these rapidly advancing areas of science.

**Box 7.2** Genomic Science as a Leading-Edge Tool against Challenges

Genomic science is prominent among the various scientific approaches to challenges for humanity and the environment, especially those concerning human health, agriculture, and biodiversity. From its origin back in the time of Mendel, who founded the science of genetics, it has gained recent major allies in molecular biology and information technology. Genetics classically aids in plant and animal breeding through selection of preferred phenotypes or physically observable traits. Molecular biology has helped in characterizing the traits in molecular details and, after the discovery of the DNA structure, down to the level of the genes. This resulted in major advances in many areas of agriculture, such as in selection of preferred varieties through selection by molecular markers, with ease in manipulation and faster and more reliable outcomes than conventional trial-and-error field studies. In medical science, molecular biology has enabled our understanding of the nature of genetic diseases and the role of genes in other diseases and their prevention. The advent of recombinant DNA technology allows genetic modification and insertion of genes from other organisms into microbes, plants, and animals, starting biotechnology industries producing biological drugs (biologics) as well as transgenic plants and animals with desired characteristics. While biologics are well accepted, the transgenic plants known as genetically modified organisms (GMOs), with properties such as pest and drought resistance, are still not accepted in many countries.

The ability to determine the long sequences of DNA, which determine the identities of the genes and their products, has increased exponentially, together with a great decrease in costs. It is now easy to determine in an organism the whole sequence of DNA that contains all the genes, known collectively as the genome. The age of genomic science was born, together with the ability to not only sequence but also analyze and predict many protein products derived from the genes and also analyze the DNA sequences that do not correspond to gene products but are involved in control of life processes. Bioinformatics, a hybrid science of information technology, molecular biology, and genetics, is now a useful tool that can predict the variety of traits from whole genomes, allowing selection and breeding of preferred varieties. New tools in molecular biology, furthermore, allow not only selection from breeding but also direct modification of the genome through techniques of gene editing, such as one using technologies of *clustered regularly interspaced short palindromic repeats* (CRISPRs). This has implications not only in agriculture but also in various aspects of health and biodiversity. For example, gene editing allows scientists to use what is called the "gene drive" mechanism to eliminate insects and other pests. The news is not all good, however. It is also possible that genetic modification of harmful bugs could be used to create biological weapons used for war and terrorism purposes.

The new technique also allows editing of various genes (caricatured in Fig. 7.2) of humans and other living organisms, with the potential to cure or prevent genetic diseases. However, it will also allow the creation of "designer babies," made from the instruction of parents or authorities. A dangerous slope has therefore been exposed, and society still has to catch up with the implications of this new technology. Bioethical considerations are of crucial importance, and the scientific community should encourage society to consider these issues broadly and together come to definitive guidelines before scientists go on relentlessly to create the brave new world, as described by novelist Aldous Huxley, in which genetic characters are determined artificially.

**Figure 7.2** "I gave my pig a gene-editing treatment."

Governments, international agencies, and professional scientific organizations should direct more efforts in mobilizing science to help alleviate societal problems. Business can help in making the tools or the generated products and services available to the public so long as they can satisfy their profit incentive in doing so. Many potential solutions to societal problems, especially problems of the poor and deprived, such as problems in health and hunger, suffer from what are called market failures. The failures are due to lack of sufficient market incentives, including profits, to make available these solutions to those who need them but either they or their governments cannot afford. Some potential solutions to the problems have already been found by science, such as many drugs and vaccines, but they cannot be accessed by the people in endemic countries or other people who need them. Programs by charitable organizations to provide these items to people in need are admirable, but these measures are mostly not sustainable in the long run because of constant reliance on financial input. New models of business, such as what are called social enterprises, which are sustainable nonprofit enterprises have recently been developed in many countries. They are part of the new potential solutions to solve the problems of lack of access to the products and other solutions that science has already helped find but that need other input to make them available to society at large.

We have noted earlier that human activities have resulted in many environmental problems. Pollution from agriculture includes released chemical fertilizers and pesticides; pollution from industry and transport includes released refrigerants, solvents, and greenhouse gases such as carbon dioxide; and pollution from household and other human activities includes release of plastic containers and various wastes. These pollutions result in environmental degradation, loss of biodiversity, and the threat of climate change. Part of these problems is due to science-led products and activities. Science therefore has the responsibility to help in the mitigation of existing problems and the prevention of future ones. The scientific approach to these problems should be able to help provide the solutions needed. For example, the problems of excess use of chemical fertilizers and pesticides can be solved by the use of new varieties, resulting from genetics-led research and integrated pest management, which are less reliant on such brute-force measures. Release of industrial pollutants can be minimized by use of new, more efficient processes that minimize such release and of agents that on release do not cause as much harm to the environment. Problems of household pollution can be minimized by waste recycling and the use of biodegradable containers and other consumer products. These are some examples of how science helps create a sustainable environment for the future.

## 7.5 Challenges in Health

Health not only means freedom from disease but also means a state of wellness of the body and the mind. Challenges to science in finding solutions to health problems range from provision of individual care to care at family, community, country, and global levels. As societies all over the world become "grayer," problems in maintaining the health of elderly people and dealing with associated chronic diseases such as hypertension, diabetes, and dementia require urgent attention. Other major problems include cancers of various types, heart diseases, obesity, and environment-related diseases. Although the root causes of many types of cancers are known, diagnosis has become more accurate, and prevention and therapy more effective than a few years ago, cancers still pose major threats to humankind all over the world. Newly emerging diseases, such as Ebola, Zika, and bird flu, will also periodically challenge us, especially as the ease of international travel carries risks of spreading these diseases. Threats of HIV/AIDS have been partly mitigated by new drug combinations, better public education, and other actions such as the prevention of infection in drug addicts. However, some 37 million people are still living with this disease worldwide. Infectious tropical diseases like malaria and dengue hemorrhagic fever are still major challenges, especially in developing tropical countries in which people are poor and live in conditions that allow the diseases to thrive with the help of insects and other vectors. Currently there are more than 200 million cases and 400,000 deaths per year from malaria [38]. Tuberculosis affects many people in both developing and developed countries, especially the poor because of bad hygiene and living conditions. There are some 10 million new tuberculosis cases each year and 1.8 million deaths, including 400,000 of those who also carry HIV [39]. Although these diseases present complex medical and scientific challenges, they should, in principle, be tamed by medical science, like many other diseases prevalent in developed countries. However, because of the problems of poverty and lack of purchasing power of the afflicted, these diseases are relatively neglected by the world at large. Moreover, global warming threatens to worsen the spread of vector-borne tropical diseases, since warmer climates are more conducive to insect vectors and disease pathogens.

In spite of these various remaining unsolved issues, medical and allied sciences have helped answer many other challenges in health, from diagnosis to prevention and therapy. These sciences make an impact not only through provision of health products but also through public education on hygiene and healthy lifestyles. For many, an advanced age need not mean lowering of the quality of life, thanks to better knowledge about health needs of the elderly and better systems for medical attention. Chronic diseases like hypertension and diabetes that pose major threats, especially to the aging population, can be dealt with much more effectively than before through new treatments and prevention. Public awareness of hazards from smoking, eating, drinking, and environmental pollution has significantly helped reduce the burden of diseases from these factors. Many long-standing ailments have been curtailed with the help of new drugs, vaccines, and diagnostics. Antibiotics have been major tools of the past few decades in the fight against infectious diseases, although we now need to deal with the problems of drug resistance in order to maintain their efficacies. Development in molecular biology and biotechnology has resulted in many biodrugs or biologics, such as blood components, hormones, vaccines, and therapeutic proteins. Unlike conventional small-molecule drugs, these are large biomolecules that work through specific molecular mechanisms at the intended target tissues. In many cases drugs are delivered to the target through the use of special systems that direct the drugs to the intended sites.

Advanced medical science that combines knowledge in bioscience with information technology (IT), nanotechnology, and cognitive science is transforming many areas, including drug development, surgical techniques, and general hospital care. This includes genomic medicine, which deals with diagnosis, prevention, and treatment of individuals through genomic analysis and the use of new tools such as gene editing (Box 7.2). Gene editing will also allow modification of genetic characters of cells and tissues so that transplantation of organs from other species to humans will be possible, obviating the need to transplant human organs. Together with the rapidly increasing ability to collect and analyze big data, genomic medicine not only leads to personalized medical care but also will soon provide new tools to improve public health from genetic knowledge of the population. A relatively new field called stratified medicine allows appropriate and effective treatment of groups classified through biomarkers. In a related development, electronic data capture allows collection of clinical and related data of individuals and groups for accurate information on treatments and other interventions, similar to checking of the financial information in business transactions. IT is also adding a new dimension to health care through telemedicine, which allows remote examination and treatment of patients. This has immense implications for people in remote communities, who cannot access health services readily. A related example of transdisciplinary development is that of robotic or computer-assisted surgery, in which a remote manipulator allows the surgeon to perform surgical procedures. Although still expensive by comparison with conventional surgery, it has potential advantages in accuracy and in reducing incision sizes, and it can be expected to gain more widespread use in the future when the cost comes down and the accuracy of the operation goes up. At the level of the lay public, cell phones and the Internet can help give information to the public on keeping healthy and being alert to health threats like emerging diseases. Sciences of robotics and intelligent manufacturing, including 3D printing, are helping disabled and elderly people through the making of artificial limbs and other devices such as aids for hearing and seeing and other organ replacements. Medical devices in general are made through collaboration between the medical and engineering professions. Hospitals and convalescing homes for the sick and the elderly have also been greatly improved through the help of new medical devices and associated technologies.

A major implication of these new trends is that health care is moving from therapy, that is, treatment after illness is diagnosed, to prevention of illness and the promotion of wellness. In many cases, the cost of health care will increase because of development of new equipment. Yet in other cases new equipment will increase efficiency and cut costs. Health care will also cost less for the public that already has tools such as cell phones for health information and simple self-care. Provided with the capability to gather and handle big data in public health records, governments and local authorities can also develop stratified medicine as an efficient and cost-effective feature of public health care.

## 7.6 Challenges in Food

Although agriculture is capable of feeding a substantial portion of the world population, one in nine people in the world still suffers from chronic hunger and undernourishment. Productivity needs to be improved drastically in view of the fact that arable land can no longer be expanded and probably will decrease owing to urbanization and population pressure. Increased population, together with urbanization, points to the need for food production to be increased by some 60% by 2050 [40]. Failure to feed the world population is due not only to a shortfall in food production but also to inefficiencies in food storage, processing, and transport to where it is needed. Inappropriate use of land, desertification, acidic or salty soil conditions, lack of water, and other factors, combined with urbanization, lead to the failure of agriculture in many places. Increased urbanization also means increased demand for meat, requiring greater amounts of feed for livestock. Foreseen climate changes will also add to crop failures, increased pests, and decreased fishery yields. Even when the raw materials have been produced, a number of factors contribute to hunger and malnutrition. It has been estimated, for example, that postharvest losses account for some 30%–40% of food produced. There is also much wastage in the food supply chain and consumption. In addition to problems in crops and meat production, aquatic products are also facing pressure. Competition for harvest of fish and other marine species is becoming intense, and it has been predicted that catches in the tropics will be 40% lower by 2050. The situation is exacerbated by illegal, unreported, and unregulated (IUU) fishing, which is threatening the ocean ecosystem because of lack of conservation and management efforts. Failures in agriculture and fishery result not only in hunger and malnourishment of the deprived but also in increased poverty of the people whose livelihood depends on food production and related activities.

These challenges must be met if humankind is to become free from hunger in the future, with the active role of sciences involved, including crop science, horticulture, animal husbandry, fisheries, and related areas. The Green Revolution around the middle of the past century (Chapter 1) helped greatly in crop production through the improvement of varieties and farming procedures, but it was limited by the requirements for fertilizer and water. Success of the early efforts therefore has to be made more sustainable through new scientific input and other measures. Genetically modified (GM) crops have been adopted in many countries and have significantly helped in the attempts for increased food production. However, the first-generation GM crops, with insertion of exotic genes, such as genes for pest or drought resistance, have not been universally accepted, mainly for fear of environmental risks. New methods, including those for genome editing, are presently being developed for plant and animal varieties, and it remains to be seen whether they will be technically successful and will be acceptable to the public at large. Looming on the horizon are farm products such as meat free from dangerous fats and vegetables rich in beneficial nutrients. However, in addition to the foreseen reward, we should also be aware of the potential risks of these new genetic techniques, which will be explored in Chapter 9.

Even without these new developments, plant genetics has already contributed to agricultural production through breeding of superior varieties assisted by the use of genetic markers. Management of farms for protection from pests and productivity can also be made through nuanced ecological approaches, such as biocontrol and the use of diverse species, rather than the crude use of pesticides and chemical fertilizers. For example, many rice farms are using the fungus *Beauveria* and other fungi for insect control. Biofertilizers like the bacterium *Rhizobium* and blue-green algae have been in use for a long time, especially by small farmers. These biocontrol agents and biofertilizers can be made by the farmers themselves through appropriate technologies, which need to be promoted both at the farm and at the industrial level. We should also explore new food sources that can be produced sustainably from the biodiverse environment. In this respect, we can learn from sustainable food traditions of indigenous people in various parts of the world and learn to adopt some of the foods in a sustainable fashion. For example, many insects, some of which are agricultural pests, are eaten by local populations, serving a double purpose of both getting nutrition and getting rid of the pests.

As for products from the sea and freshwater sources, we should concentrate more on the development of aquaculture of fish, shrimps, and other aquatic products. We should look forward to the age of farming, not hunting, for aquatic products. Recent advances in aquaculture have made intensive farming of many kinds of aquatic animals possible, although the hazards of viral and other diseases that threaten to wipe out whole stocks still have not been adequately addressed, especially in small-scale open aquaculture. Closed-system aquaculture is a safer farming method, but this is only possible with big operators and not small farmers. Rapid diagnostics are needed to detect viral and other infectious agents that can wipe out the whole farm quickly. We should also eat more aquatic plants such as seaweeds, as is already done extensively in many cultures. Many seaweeds are relatively easy to grow and harvest by small farmers and fisherfolk.

In addition to agricultural sciences leading to the production of raw materials, food production also depends on related technologies, including those dealing with postharvest collection, preservation, processing, and distribution. The science of nutrition, dealing with our dietary requirements, food quality, and safety, is also important for keeping our health in relation to food intake. The food industries can meet these challenges with the aid of bioscience, food technology, packaging, and other technologies. For example, farm produce can be kept fresh over long periods through packaging with appropriate materials. Drying and refrigeration can help preserve fruits and meat, and the technologies can range from simple sun drying at the household level to industrial-scale refrigeration. Traceability of products from the farm to the market can help build the confidence of the consumers in the quality of the materials. Food safety and quality standards need to be ensured by the producers, and monitored by appropriate authorities, both at the premarket and at the market level. These measures are important, especially for exported and imported foods, not only for the consumers, but also for the livelihood of workers in the food industries.

Finally, for the consumers, who are at the end of the food supply chain, awareness of nutritional values of food is important, especially for vulnerable groups of people, including children, the ill, and the elderly. Obesity and related ailments, including diabetes, heart disease, and stroke, are significant health problems resulting from unsuitable food, excessive consumption, and physical inactivity. According to the World Health Organization (WHO), worldwide obesity has more than doubled since 1980, and 13% of adults are now obese. Obesity is now killing three times more people than malnutrition. Problems related to obesity are the flip side of hunger and malnutrition, all of which are obstacles to sustainable development.

## 7.7 Challenges in Living and for the Environment

If people from 200 years ago could be brought back to life, they would hardly recognize the world around them. They would be awestruck by modern utilities in homes and be amazed by cities full of skyscrapers. The means of transport, with airplanes, cars, and the mass transit system, were barely as predicted from science fiction of the time. Not only have the physical features of the living changed beyond recognition and dreams, but so have the ways of life. Our jobs, how we go about our daily lives, and how we communicate with one another have also been drastically transformed. All these changes did not come without cost. Although there have been increased efficiencies in how we use raw materials and energy, consumption of these resources went up with the changes in living and the increasing population. We may think that compared to 200 years ago, at least nature and the wilderness have not changed much. That is an illusion: our forest cover has dwindled together with its biodiversity, our water resources are reaching critical levels, and we have burned up major supplies of fossil fuels in just a hundred years or so. We have polluted our planet both on land and in the oceans with industrial and household wastes. Greenhouse gases, including carbon dioxide from fuel combustion, methane from agricultural activities, and noxious fumes from industry and transport, all add to atmospheric pollution and climate change, by far the biggest problem of our and future generations.

These are major challenges that need to be met, as recognized by the UN General Assembly, which has adopted 17 Sustainable

Development Goals to be met by 2030. We will discuss these issues in full in Chapter 10. Here, we will examine the role of science in helping answer some of these challenges. Increased connectivity, enabled by the Internet and other information and communication technologies, has transformed the nature of work for many people and will continue to do so at a gathering pace. This, together with advances in robotics, will eliminate many jobs from manufacturing and service industries. However, this transformation should also lead to the creation of new jobs such as those related to production and servicing in cell telephony, intelligent devices for consumers and industries, financial technologies, and new ways of servicing the public. The Uber taxi type of services, jobs related to online shopping, and freelance jobs concerned with the use of various skills, including data entry, accounting, and graphic design, are some examples of the new job landscape. The demographic shift toward aging societies in many countries is creating more jobs concerning care of the elderly and maintenance of good health. Conventional jobs can also be done more at home, reducing travel time and energy costs. It is difficult to predict whether the new job situation will help reduce social inequality and offer more opportunities to women, but the fact that many jobs will be done more at home rather than in workplaces should help in tipping the balance toward women.

The energy picture is changing rapidly. Renewable sources such as solar and wind energies, regarded only a few years ago as nonconventional, are now on the verge of becoming conventional, replacing fossil fuels with their associated problems. These changes are having major effects on transport, industry, and household energy usage. Advances in energy storage and battery technologies now make solar roofs desirable for many homes, and the excess energy from these can be supplied to the grid system, reversing the conventional model of one-sided reliance of the homes on the grid. With ease of charging, either from the homes or from gas stations, electric cars could replace conventional fuel-driven cars in the not-too-distant future. Vehicles using hydrogen as fuel are also being developed as alternatives to fossil fuels. Although these produce water as the result of energy use rather than greenhouse and polluting gases, the methods of producing hydrogen that cost energy are still major problems that need to be overcome.

In contrast to the energy outlook, the water situation is still unclear, except for the certainty that there will be far increased demand in the future. The two major problems that need to be solved are supply of water and its quality. Competing demands for water from agriculture, industries, and household consumers will put increasing pressure on society and cause major shifts in land use and community locations. Large dams are not the preferred solution, since past experience shows that they caused hardships to people due to relocation of upstream people, depletion of water for downstream people, increase of some water-borne diseases, and other problems. Dams also contribute to climate change through change of water flows and storage patterns. It is more preferable to make small-scale dams for use by communities. There will be increasing pressure to use underground water, which is a major source but plays a part in land stabilization. Science must help meet the need to explore deep underground water and its use without causing land subsidence and other unwanted side effects. Desalination of seawater through reverse osmosis and other techniques should also be developed further with a view to decreasing the cost and energy required. With regard to supply of clean water, technologies are mostly in place for big cities with facilities for treatment. However, low-cost solutions are needed for people in developing countries to obtain potable water at the village level in technologically backward, remote locations far from urban centers. These may involve the use of safe chemicals for disinfection, low-cost filters, water-purifying pumps, and solar sterilization. Often, these technologies are simple and their effectiveness proven on a small scale, but they are difficult to scale up for mass markets. Such scale-up may need a model of business like what is called social enterprise, which does not aim for profit but for social return and is backed by government policy such as tax exemption.

## 7.8 Challenges in Sustainable Production and Consumption

Sustainable living in harmony with the environment requires that the world have sustainable, or green, production and consumption. At the global level, production and consumption critically affect processes such as climate change and other human effects on the biosphere. The land and oceans must be continuously monitored and appropriate actions taken at the international level. At the country level, production and consumption affect economic growth and social and environmental status. Governments must respond by policy and action for issues such as waste minimization and recycling, utilities and energy supply, the built environment, land use, and nature conservation. At the level of the community, the family, and the home, production and consumption affect our lives in how we eat, work, and entertain ourselves and in various other things we do in our daily lives.

Science and allied subjects are key actors in meeting the challenges of sustainable production and consumption. In sustainable production, raw materials and processes must be optimized so as to achieve not only economic profits but also minimal costs to the environment and the society as a whole. This demands judicious choice of input, not only in materials, but also in energy, human labor, and other factors such as machineries, water, and transportation. The production should minimize waste materials, which must be subject to management and recycling. It should cause minimal or no disturbances in producing noise or in releasing gases, dust, and smell. The production of greenhouse gases is one measure of whether sustainable production has been achieved. Furthermore, the task of sustainable production does not stop merely at manufacture but goes on throughout the life cycle of the product. How much greenhouse and noxious gases will be produced in the lifetime of a car? How much heat and coolant will be released by an air conditioner? How will the food packages be disposed of, and will they obstruct waterways or pollute the oceans? At the end of its useful life, can the product be recycled? These are some of the issues that science and engineering must deal with in collaboration with allied subjects in social, economic, and management areas. They have been successful in many areas and have helped manufacturers design product life cycles through the zero-waste philosophy, with three Rs, namely *reduce*, *reuse*, and *recycle*, as a basic principle. They have also in many cases advanced from the concept of "cradle to grave" for products, which must end up in landfills or become burdens to the environment, to that of "cradle to cradle," where products such as plastic bottles can start new lives from recycling. Recycling of parts of cell phones and computers is more difficult, but e-waste management is a field that is developing rapidly as part of sustainable production.

The 1994 Oslo Symposium on Sustainable Consumption defines sustainable consumption as "the use of services and related products that respond to basic needs and bring a better quality of life, while minimizing the use of natural resources and toxic materials as well as emissions of waste and pollutants over the life cycle of the service or product so as not to jeopardize the needs of future generations." It requires wise and efficient use of resources, with minimization of waste and pollution. Necessarily, this is closely linked to sustainable production but puts a major responsibility on the consumers. In promoting sustainable consumption, awareness of the public of its importance has to be built. Public education is important, but the role of social media and mass media is crucial in this effort. The public should be able to gain information from the Internet and other sources. People will realize that sustainable consumption is not only good for the environment but also good for their own pockets. Having smart homes, with smart designs and building materials, using efficient light sources and energy-saving devices, eating moderately, and minimizing household wastes all add up to sustainable consumption. The government, local authorities, and communities have major roles in encouraging sustainable consumption through policy and services that encourage recycling, waste minimization, and waste separation. Although the public agrees in general with the concept of sustainable consumption, it is not easy to translate this into routine action in daily life. However, efforts must continue to be made, taking lessons from countries and places that have been successful in adapting sustainable consumption to the local situation.

# Chapter 8

# Addressing the Base of the Pyramid

*The problem of poverty must force us to innovate . . .*
—C. K. Prahalad, *The Fortune at the Bottom of the Pyramid* (2004)

The base of the social pyramid comprises people with the lowest income and earning power. It should be addressed by the whole society, not only because of political, economic, and humanitarian reasons, but also, more importantly, for the integrity of society, which depends on all members. Poverty, diseases, and lack of education and opportunities, all combine to hamper the rise of the masses of people otherwise equal in potential to the luckier ones higher up in the pyramid. With the help of education and other social programs, science can solve problems of the base of the pyramid and help it achieve its potential. The base of the pyramid, with untapped potential due to its myriad problems, indeed offers huge potential markets for products and services in many areas, including food, water, health, and small industries. Meeting these demands will lead not only to a more equitable and just society but also to a higher level of prosperity overall because of bigger contributions from the base. Science is an important component for meeting the needs of and for the development of people at the base of the pyramid.

*Sparks from the Spirit: From Science to Innovation, Development, and Sustainability*
Yongyuth Yuthavong

ISBN 978-981-4774-57-4 (Hardcover), 978-1-315-14599-0 (eBook)
www.panstanford.com

## 8.1 The Base of the Pyramid: A Latent Power

If people are grouped by their wealth, the different tiers would form the structure of a pyramid. The top, comprising the very rich, is a tiny tip, under which the increasingly larger lower tiers lie. The base of the pyramid contains the largest number of people, with very low income and little purchasing power (Fig. 8.1). These people are mostly located in Asia, Africa, Latin America, Eastern Europe, and many island states. Because of its size, the base of the pyramid is important for the well-being of countries, and governments struggle to develop policies and actions to address its problems. It has been estimated that this base, where a person earns less than $3000 per year (in local purchasing power), consists of 4 billion people out of a total world population of some 7 billion. From one angle, this is a big burden for countries to take care of, especially for developing countries, where most of the people at the base of the pyramid reside. Yet, from another angle, this represents a big opportunity for countries to tap into and achieve both economic development and social equity in one go. The first viewpoint, a conventional one until recently, takes people at the bottom of the pyramid to be those who can hardly help themselves and have to wait for help from those at higher tiers through charity or social welfare. The second viewpoint takes these people to be those with as much potential as anyone in the pyramid but who simply lack the means and opportunities to carry out work and earn income that would free them from poverty. It also takes account of the fact that the sheer number of these people offers a vast potential market for both them and others to develop for the benefit of everyone concerned. This viewpoint received initial impetus from corporate strategists like C. K. Prahalad and S. Hart [41], who pointed out the potentials of business for and by the people at the base of the pyramid.

The needs of people at the base of the pyramid center around food sources and livelihoods, housing, energy, water, health, transportation, and information and communication technology [42]. Since the majority of these people live in rural areas with poor infrastructure like roads and electricity supply, they are already at a disadvantage in trying to meet their needs. Many products and services cost more and are difficult to access, giving them a further penalty. However, other factors operate in favor of markets for the poor so long as the products and services offered are appropriate to their needs. In food production, for example, small enterprises can make good business from the fact that the raw agricultural materials are produced right in the rural areas, where the base of the pyramid lives. Big companies can also profitably target markets at the base by concentrating on products such as detergents, household appliances, and low-cost cell phones. Success depends on having good distribution networks and tailoring the products to the small buying power of the people. They can also source their raw materials such as dairy supplies, fruits, and crops from the rural areas, where the base of the pyramid lives. Cooperatives for production of farm products, together with facilities for small loans and investments, can help strengthen the capabilities of people at the base of the pyramid. Finance for some of these products as well as for other small operations at the community level can also be obtained from microfinance schemes operating in many countries. One example of such schemes is offered by Grameen Bank in Bangladesh, pioneered by Muhammad Yunus, who won the Nobel Peace Prize for microcredit schemes for the poor. At the household level, people can adopt a sufficient style of living, not entirely dependent on earning cash, growing, and making their own foods and selling the surplus to markets in the community.

**Figure 8.1** A lopsided pyramid of people: a small number at the top, with large incomes and many opportunities, and large numbers at the base, with small incomes and poor opportunities.

Science can play a big role in improving the lives of people at the base of the pyramid both in rural and in urban areas. Health and food are two main areas where the contribution of science is crucial, as discussed in more detail later. Mobile and Internet technologies are helping people in making connectivities beneficial for their livelihood, health, and social activities. Access to clean water is still a problem in many areas, but there are many new, low-cost appliances such as filters (ceramic, charcoal, and sand filters), solar sterilizers (such as plastic water holders in sunlight), solar distillers, and low-cost, safe chemicals. Solar energy is becoming rapidly more affordable and available in small-scale settings. Together with development at the grassroots level is the increased reach of services in roads, electricity, and water supplies by the central and provincial authorities. The foreseen increase in the share of renewable sources in energy supply, including solar energy and bioenergy, is good news for the base of the pyramid, provided that the central governments develop appropriate policies and actions. In the area of housing, the science of materials and construction can help people at the base of the pyramid build their own houses, or substantially reduce the cost through good selection of materials. The potential contributions of science to the quality of life at the base of the pyramid are briefly discussed in Box 8.1.

**Box 8.1** Science for People at the Base of the Pyramid

The needs of people at the base of the pyramid mostly concern essential products and services. It has been estimated [42] that, by far, the largest market centers around food products, followed by energy, housing, transportation, health, information and communication technologies, and water. A large portion of these needs is met by purchase of the products and services from the market provided by large industries. However, significant portions of these needs are met by small producers and service providers at the community level and by the people themselves. Many people rely on their own food production from their farms, and the surplus can be further sold in the market. This is the basis for a sufficiency economy [43], pioneered by the late King Bhumibol Adulyadej of Thailand, which will be discussed in more detail in Chapter 10. Production of food, energy, and other items for everyday needs relies on basic skills, a large part of which can be provided by scientific knowledge at the grassroots level. Some examples of science at the base of the pyramid are:

- Health: Prevention and therapy of neglected diseases (such as tropical infectious diseases) and self-care through nutrition and hygiene, sex and general health education, herbal remedies, avoidance of mosquitoes, and other disease-carrying pests
- Food and agriculture: Basic knowledge, for both consumption and livelihood, about plants and animals, varieties, soil, fertilizers, feeds, irrigation, pests and diseases, agricultural machineries, and markets for various products
- Water for household use: Filtration, flocculation, disinfection, sterilization, and other techniques for obtaining clean, potable water
- Energy: Biofuels, including biodiesel, and other sources of renewable energy, energy from agricultural and household wastes, solar heaters, dryers, and electric panels
- Housing: Construction materials from natural sources, brick making, furniture and other household materials, and flood-resistant and waterproof materials
- Transportation: Land vehicles, car maintenance, boats, and other means of transportation
- Information and communication technologies: Use of the Internet and computers for education, communication, and livelihood (including accounting, planning, and record keeping)

## 8.2 Too Poor to Afford Health?

Access to health care is one of the most important basic human rights. People at the base of the pyramid have an increased burden in taking care of their health compared to those in higher tiers. Like those in higher tiers, they are faced with health problems centering around noncommunicable and chronic diseases and problems arising from unhealthy styles of living, such as smoking, drinking, and eating unhealthy foods. In addition, people at the base who live in remote rural areas are exposed to tropical infectious diseases like malaria, schistosomiasis, and trypanosomiasis. People at the base of the pyramid who live in urban areas tend to have a greater incidence of tuberculosis, HIV/AIDS, and other diseases that are linked to poverty. Their children are more likely than those of higher tiers to die from respiratory infections, diarrhea, and other diseases. Malnutrition, another health problem related to poverty, is common for people at the base of the pyramid wherever they live. Diseases that disproportionately affect the people at the base of the pyramid are relatively neglected, with few effective medicines and vaccines, and are addressed by only a small portion of global spending on health research and development (R&D). Since most of the people affected by these diseases cannot afford treatment costs, and public health measures in these countries are inadequate, there is a need to develop mechanisms to support R&D on these diseases. Pharmaceutical companies are not attracted to investing in the development of drugs and vaccines against these diseases, since the people under threat and their governments do not have the resources required to buy the products. In short, the therapy and protection of these "neglected diseases" suffer from market failures—inability of the normal market mechanism to serve the needs of the poor because of an inadequate return on investment.

Despite the failure of market forces, there are reasons to be hopeful. First, many developing countries have the potential capacity to themselves develop and produce drugs and vaccines. However, to be able to do so, they will likely need assistance from international communities and need to exert much more effort than they currently do in meeting internationally accepted quality control standards for good manufacturing, laboratory, and clinical practices. Second, there is a growing realization that it is in the interest of people in rich and poor countries alike to be protected from the threat of emerging diseases, new and old, that are being spread among the rich and poor alike through increasingly fast and dense networks of global travel and trade. Therefore, this problem should attract the attention of public sector decision makers all over the world. Third, many pharmaceutical companies are becoming more interested in the development of drugs for diseases that predominantly affect the poor. One reason is the realization that they have the social responsibility, not only to their own customers and share owners, but also to all the people of the world, including those at the base of the pyramid. Another reason is that in this increasingly connected world, problems are not confined only to people at the base of the pyramid, or to any regions of the world, but have the potential to spread to other people in all parts of the world. Lessons from the spread of emerging diseases like Ebola, Zika, and severe acute respiratory syndrome (SARS) bring home the realization that we as humans are under common threats of diseases, which are no longer only confined to people at the base of the pyramid, and we have to find ways to prevent and cure them together.

World attention on diseases that predominantly affect the poor was reflected in the UN Millennium Development Goals, which called on countries to combat HIV/AIDS, malaria, tuberculosis, and other diseases by 2015. These goals were partly successful and are now taken up further as parts of the Sustainable Development Goals launched in 2015 to ensure healthy lives and promote well-being for all at all ages. Some problems, such as prevention of disease transmission by mosquitoes, have been successfully tackled through widespread use of insecticide-treated bed nets, a solution brought about by both scientific research and large-scale production through good management practices. Diarrhea is effectively treated with a low-cost oral rehydration formula developed from research in Bangladesh and other places. Other problems, including development of effective vaccines and therapies, still need more scientific R&D.

A part of the world community of researchers has been drawn together to fight neglected diseases over a number of years. Under the auspices of the World Health Organization (WHO) and other organizations, the Special Programme for Research and Training in Tropical Diseases was formed during the 1970s. Around the same time, the Rockefeller Foundation launched the Network on Great Neglected Diseases of Mankind, with participation of a number of scientists from distinguished universities around the world. The beginning of this century saw the launch of the Grand Challenges in Global Health Program, initiated and funded by the Bill and Melinda Gates Foundation, dedicated to supporting scientists and innovators in finding new solutions to the problems of health and diseases, especially those affecting people at the base of the pyramid. Box 8.2 gives an account of some scientific efforts dedicated to answering the challenges of neglected diseases. The problems are far from being solved, but at least recent examples of the threats of emerging diseases have energized the world into much greater action than before.

**Box 8.2** Challenges of Neglected Diseases

People at the base of the pyramid are mostly powerless to solve their problems concerning diseases that predominantly affect them, both through the fact that many of these diseases are associated with their living conditions (insects, unclean water, poor hygiene) and through the fact that prevention and therapy are not available to them. In addition, many diseases still require scientific R&D of the tools to fight them. The efficacy of vaccines against malaria, for example, still needs to be improved, as also drugs against the parasites that can rapidly evolve drug resistance. The treatment of tuberculosis is arduous owing to the multiple drug regimens that are required to overcome drug-resistant pathogens. There are still few drugs against diseases such as schistosomiasis, sleeping sickness, and many viral diseases. Diseases such as hepatitis C, which affect both people at the base of the pyramid and people in higher tiers, already have effective drugs, but they are too expensive for the poor.

International scientific communities are rising up to the challenges in fighting neglected diseases. The 1970s saw the establishment of the Special Programme for Research and Training in Tropical Diseases, which concentrated on diseases like malaria, schistosomiasis, sleeping sickness, Chagas disease, elephantiasis, and leprosy. The Rockefeller Foundation started the Network on Great Neglected Diseases of Mankind, under the leadership of Ken Warren, which established 14 research units of excellence in various parts of the world, including developing countries, to tackle these problems. The neglected diseases were for the first time highlighted as challenges of newly rising biomedical science, until then mostly preoccupied with diseases mostly affecting people in the higher tiers of the pyramid. A great boost to the science of neglected diseases came in 2003, when Bill and Melinda Gates launched the Grand Challenges in Global Health Program to address health problems of the majority of the world who are at the base of the pyramid. The term “grand challenges” was used in reference to the words that the great mathematician David Hilbert used in 1900 to challenge mathematicians to solve 23 major problems of the time. For the new century, major challenges are directed toward global health, with 14 challenges covering 7 goals, namely developing improved vaccines, developing new vaccines, controlling insect vectors, developing nutritious crops, limiting drug resistance, curing chronic and latent infections, and measuring health status. Some of these challenges have been successfully addressed, such as development of nutritious varieties of crops and fruits, new vaccines, and genetic means to reduce mosquito populations [11]. The “grand challenges” concept for support of research to address problems of the poor and the underprivileged and for development in general has been well received in many countries and has led to the formation of “grand challenges” programs in other countries, including Canada, Brazil, South Africa, Korea, and Thailand.

## 8.3 Too Poor to Feed Themselves?

We have already seen in Chapter 7 that food production presents many challenges, from raw material production through postharvest management, food processing, and transport to markets until it finally reaches the consumers. Consumers in the higher tiers of the pyramid with the financial means to obtain food have relatively fewer problems, mainly on food quality and safety. Those at the base of the pyramid have far greater problems: they have few financial resources to pay for the food or are at the margin of cash-based economies and have to grow or find their own food. In good years they can grow their own crops or tend to their own animals, surviving on their own outputs and selling the surplus to the markets. In bad years, however, they have to go hungry. People in Ethiopia, Sudan, and other countries in Africa and South Asia periodically suffer years of famine, which may be related to climate change and other poorly understood global conditions. Bad government policy and political turmoil also contribute to famine in many countries. While famine, which brings death and starvation, is occasional in occurrence, more people suffer from chronic hunger and malnutrition. This results from poor agricultural yields, in turn resulting from poor seed or animal varieties, poor soil, lack of water, pests and diseases, lack of feeds for animals, and other causes. People whose livelihoods depend on aquatic sources also suffer from dwindling catches from the rivers or the seas, and most of them are too poor to invest in aquaculture, which is a more sustainable and economically viable means to obtain aquatic products. Even farmers and other people in higher tiers of the pyramid, who make a living from selling their agricultural products, have problems with poor yields, low prices, and other factors. For example, when the harvest is bad, there is little product to sell, but when the harvest is good, prices tend to drop. Developed countries like Japan, the United States, and those in Europe have state policies and subsidies to cushion them from these problems, but developing countries usually have to fend for themselves without such help.

International and philanthropic organizations, together with many governments, have exerted considerable efforts to help farmers produce enough food and to free the world from hunger. A major example is the Consultative Group on International Agricultural Research (CGIAR), launched in 1943 by the Rockefeller Foundation and the Mexican government and later joined by many other funding agencies, governments, and international organizations. The consortium supports a number of international research centers working on production of important crops, including wheat, rice, and cassava, and on other problems of tropical agriculture. The strategic aims are not only to produce more and better crops but also, broadly, to reduce hunger, ensure food security, improve health and nutrition, and sustainably manage natural resources. A more detailed account of the work of the CGIAR is given in Box 8.3.

**Box 8.3** Consortium on International Agricultural Research

Agriculture is important for people at the base of the pyramid, both to free themselves from hunger and to earn from its output. Agriculture in developing countries faces many problems ranging from poor soil conditions, lack of water, and pests to low prices of the products. Among many attempts by international agencies and governments to solve these problems, the consortium of the CGIAR stands out as the most important one. It was first launched in 1943 through joint efforts of the Rockefeller Foundation and the Mexican government and later established in 1971 as a global partnership of governments, philanthropic foundations (including the Ford and Rockefeller Foundations), and various international institutions, including the Food and Agriculture Organization (FAO) of the United Nations, the International Fund for Agricultural Development (IFAD), the United Nations Development Programme (UNDP), the World Bank, the European Commission, the Asian Development Bank, the African Development Bank, and the Fund of the Organization of the Petroleum Exporting Countries (OPEC Fund). It is dedicated to research with the objectives of reducing rural poverty, increasing food security, improving human health and nutrition, and ensuring sustainable management of natural resources. It laid the seeds for the green revolution in the middle of the past century, with the development of high-yielding, disease-resistant varieties of cereals that helped turn countries like India that were facing starvation to becoming net exporters.

From an initial focus on improvement of staple grains such as wheat and rice, the CGIAR, now with 15 research centers, has evolved to include research on crops, livestock, fisheries, forestry, and management of natural resources, improving agricultural policies and market access and strengthening institutions. This has produced a substantial improvement in the livelihoods of the poor in developing countries. The vision of the CGIAR is "to reduce poverty and hunger, improve human health and nutrition, and enhance ecosystem resilience through high-quality international agricultural research, partnership and leadership" with three strategic objectives, namely "Food for People, Environment for People, and Policies for People." It now has a total of more than 2000 scientists working in 100 countries, with more than $500 million invested each year for R&D in agriculture, primarily for the poor in developing countries.

Problems concerning food and agriculture are linked to many other problems on sustainability, especially for people at the base of the pyramid. Lack of food or income from farm and food products leads to hunger and poverty. Poor health and poor nutrition result directly from these problems. As vulnerable members of society, women and children tend to suffer more. Children cannot develop normally, both physically and mentally, and mothers cannot take proper care of children, leading to a poor outlook for the next generation. Goals for sustainability, such as slowing undesirable climate change and conserving biodiversity, are adversely affected, as are other important goals for economic and social development. It is therefore important to look at problems of food and agriculture, as also other problems in development, not as separate problems but as being interconnected and requiring efforts together on many fronts.

## 8.4 Small Is Beautiful and Appropriate Technology

Development of modern industries and other commercial activities tends to be associated with large-scale production and services, which bring in good returns due to the economies of scale, with lower product and service costs, and other factors such as market size. However, such activities often bypass people at the base of the pyramid, except recruiting some of them as manual workers with low wages. Technology, which is an important component of the production and services in modern industries, is seen by many at the base of the pyramid as inhuman and as playing a role in reducing jobs for people because of the automation that it brings. The past few decades that saw massive changes in production and services brought about by automation and other technologies did not see large-scale unemployment, as might have been anticipated, due to the birth of new jobs in services and other sectors of the economy. However, the accusation still stands that in many cases, technology lacks a "human face." In the 1970s, a movement was initiated to make technology more friendly, especially to people at the base of the pyramid. E. F. Schumacher [32] called this "intermediate technology," as its purpose is to help the common people in their livelihoods and daily lives. Intermediate technology, or appropriate technology as it is also called, is "technology with a human face," suitable for small-scale production and services, and is part of the "small is beautiful" movement, which proposes a model of economic development that puts importance on the integrity of people. Examples include simple water pumps, water filters, solar cookers, solar lamps, building materials from local sources, biofertilizers, and biopesticides. Many of the technologies are generated by the people who use them and therefore know what is needed from personal experience. The technologies are characterized by their ease of access by people, their moderate costs, and the possibility of improvement by the people who themselves use the technologies.

Promising though the movement of appropriate technology is for the common people, especially those at the base of the pyramid, it fell into disfavor with some who associate it with second-rate technology. Reliance on home-grown technologies, while responding to what is needed at the base, suffers from disconnect with advances in science and technology, many of which are mostly associated with mainstream industries and services and protected by law as intellectual property. In reality, as the name suggests, "appropriate technology" is the most suitable technology in the context of common users who are not involved in large industrial applications and yet want tools made available by technology that they can handle themselves. People at the base of the pyramid are among the most important groups to benefit from the movement of appropriate technology.

The age of sustainable development that we are trying to build should bring in new realizations about the importance of appropriate technology in the everyday lives of the common people, including those at the base of the pyramid. The charm of appropriate technology is the fact that it is small scale and people oriented, incorporating elements such as local culture and originality of individuals and the community. For example, household goods, cultural souvenirs, and local food items can be made through appropriate technology, with good quality control. Appropriate choices of products and clever marketing strategies, such as in the One-Village-One-Product scheme of Japan and the similar One-Tambon-One-Product scheme of Thailand, can also help in the success of small-scale, people-oriented industries.

The revival of interest in appropriate technology can utilize new connectivities enabled by information and communication technologies, such as social networks and online learning. It can also be nature friendly, utilizing materials from nature in a sustainable way through processes that are not energy intensive and do not release excessive greenhouse gases. Appropriate technology is therefore not only people friendly but also environment friendly and can become pioneer sustainable technology in the post–industrial age period. New trends in innovation in areas such as biotechnology and information technology, which favor openness as opposed to secrecy and crowd sourcing as opposed to individual inventiveness, also favor revival of appropriate technology in the "small is beautiful" sense.

## 8.5 Science for the Base of the Pyramid

Problems for people at the base of the pyramid require innovative solutions in the sense that they must be simple, affordable, and sustainable. Science is a main potential contributor to these solutions, as highlighted in Box 8.1. However, they require special approaches distinct from normal ones used to serve industries and services, which bring in foreseen financial returns and therefore can attract the investment needed. People at the base of the pyramid cannot afford to pay high prices for products and services. Many of them also live in rural and remote areas, which lack basic facilities such as clean tap water and electricity. Solutions to the problems of people at the base therefore require special scientific approaches distinct from normal ones. The search for such solutions requires not only research of high quality but also development, including various partners who together can make the products and services available to the users. Simplified diagrams for the processes of making and selling products for people at the base of the pyramid are given in Fig. 8.2. In many cases (Fig. 8.2a), a normal market mechanism works, where product makers and service providers, with suitable input from investors, can make their products and services available for users through purchase from sellers. These could include consumer products like shampoos, packaged foods, or cell phones or services like local transport. The role of science is to help in making or improving products and services, reducing the cost, improving quality, and making the products and services suitable for a local context. Although individuals lack buying power, the strength of the base of the pyramid is the large number of potential customers. Appropriate adjustments of raw materials and production processes, or of product packaging to suit the consumers' preference, could provide the key to the fortune at the bottom of the pyramid. A bigger role of science, however, is in introducing new products and services to mass markets at the base of the pyramid. The penetration of mobile telephony and the Internet to the market at the base of the pyramid, for example, is not only helpful in business but also provides revolutionary means for social networking, health care, livelihood, and education.

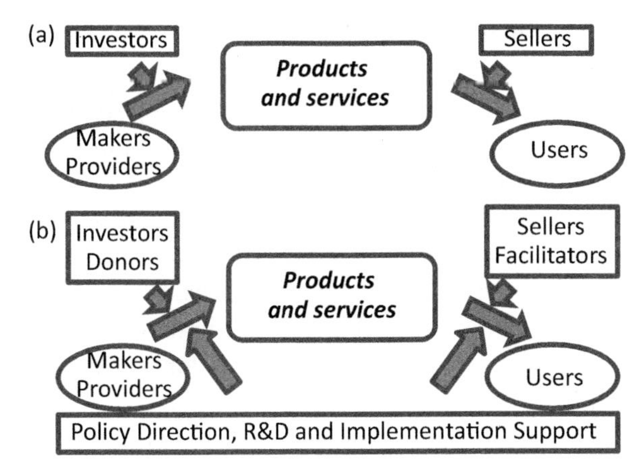


**Figure 8.2** Products and services for people at the base of the pyramid can be made through normal mechanisms as in (a) where product makers and service providers, with resources from investors, make products and services available for selling to the users or in (b) where normal market mechanisms fail, such as provision of vaccines and social services, and where donors also play a part in the investment and facilitators play a part, in addition to sellers, to make products and services available for the users.

As we saw in the examples of health, agriculture, and social services, the normal market mechanism fails the people at the base of the pyramid. They are not able to buy products and services that they need, without the active support and participation of actors other than product makers, service providers, investors, and sellers. The products and services may only be made available through R&D efforts and other policy and support mechanisms from the government and local authorities. Protection from emerging infectious diseases, for example, would not be possible without active policy support of government and international communities. When a vaccine is available, it is likely the cost would need to be borne not just by investors but also by governments and donors, who can help support its development, production, and distribution. In addition to commercial sellers, the government and local authorities often play a large role in making products and services available to the base of the pyramid. This role is shown in Fig. 8.2b. In some cases, the products and services can be made available to consumers with subsidies from philanthropic donors, and in other cases, they are provided as part of social security services or universal health care. The health-associated products include insecticide-treated bed nets that help reduce the burden of diseases transmitted by mosquitoes, the oral rehydration formula that helps in the treatment of diarrhea, and water-purifying tablets that contain flocculants, disinfectants, and chemicals that kill germs in the water. Agricultural products include biofertilizers such as nitrogen-fixing bacteria and biopesticides, both as a complete formula and as starting materials for farmers to cultivate and use by themselves. Many of the products and services for people at the base of the pyramid are best made by the people themselves, for their own use or for markets in the local communities.

Science for people at the base of the pyramid is ideally part of the knowledge and skills for healthy living and for making livelihoods suited for people of low income. However, it is also suitable for everyone who prefers the do-it-yourself and self-reliant style of living, irrespective of income status. A good education system, instilling basic scientific principles and appreciation of the ability of individuals to accomplish various tasks by themselves, should go a long way toward helping people at the base of the pyramid as well as everyone in society in general to live full and healthy lives.

# Chapter 9

# Dangers and Risks in the Sparks

> *In recent years, science and technology has progressed very rapidly and brought tremendous benefits to our daily lives. On the other hand, advances in S & T have brought about problems such as global warming, ethical concerns in biosciences and information security issues in ICT. These are the so-called "lights and shadows of science and technology."*
>
> —Koji Omi, Opening Ceremony Speech at the 4th Annual Meeting of the STS Forum, Kyoto, 2007

Science has been greatly beneficial to humankind in upgrading the quality of life and promoting social and economic development through its products and processes. However, outputs from science have also been used in harmful and unethical manners, such as in warfare, terrorism, and digital crimes. Some outputs have unintended side effects, such as unforeseen damage to health or the environment. There is a need for ethical, social, and legal considerations for science and technology. Risks from products and projects stemming from science and technology should be thoroughly assessed and preventive and mitigation measures undertaken to minimize the risks. Examples are provided for issues that concern risks and their prevention and mitigation in various areas of science and technology. Another issue sometimes encountered in science is false claims on discoveries, which require the vigilance of the scientific community to guard against. There is a need for good

*Sparks from the Spirit: From Science to Innovation, Development, and Sustainability*
Yongyuth Yuthavong

ISBN 978-981-4774-57-4 (Hardcover), 978-1-315-14599-0 (eBook)
www.panstanford.com

governance in science, which should be not only under the purview of the scientists themselves but also with involvement of the public and professional people concerned with the issues of broad aspects of science in society.

## 9.1 Light and Shadows of Science and Technology

Science is powerful. Knowledge about fission of heavy atoms led to the development of nuclear energy, which has been used both in peace and in war. Knowledge about genes and stem cells can be used in therapy and health promotion, but it can also be used in unethical ways, such as production of biological weapons containing harmful germs and biomaterials. Chemistry has given us a myriad of useful materials in our daily lives, but it can also be used to produce toxic chemicals and chemical weapons. Scientists and technologists develop useful and powerful things and gadgets, with the expectation that they would be used for various benefits. However, history has shown many times that the powerful knowledge of science was used in ways that harmed humans and the earth, such as in making weapons of mass destruction and in endangering the environment. In the words of Koji Omi, a Japanese politician who started the annual Science and Technology in Society forum, science and technology have brought on not only light in terms of benefits to humankind but also shadows, ranging from global warming to threats from biological and information technologies. It might be argued that the use of science and technology in harmful and counterproductive ways was the result of political and other directives, not from the initiatives of scientists and technologists. Such denial of responsibility may have some justification, but it cannot be accepted completely.

As a case study, we can re-examine the birth of the Manhattan Project, which gave rise to the atomic bombs that the United States dropped on Hiroshima and Nagasaki, leading to the surrender of Japan and the end of the Second World War. Leo Szilard and Albert Einstein wrote letters to President Franklin D. Roosevelt during the war, warning him that Germany might develop atomic bombs and that the United States should start its own nuclear program.

Roosevelt responded positively, giving birth to the successful Manhattan Project. Einstein later regretted his action, saying that had he known that Germany would not succeed in developing the atomic bomb, he would have done nothing. The fact, however, is that nuclear science was used to make weapons of mass destruction, which were actually employed as a means to end the war. Hundreds of thousands of human lives were lost in the bombing, and many lives were endangered as a result of the radiation long after the event. It can be asserted that the bombs led to the end of the war, but there is no denial that huge collateral damage was done.

The fact that science and technology have been used to cause immense damage to human beings and the earth has prompted comparison with powerful monsters. Godzilla is a giant monster in films from Japan, first made by Ishiro Honda, which was awakened and empowered by nuclear radiation. It caused widespread damage and defied many attempts to contain or destroy it. The question posed by these films is whether Godzilla was created or rejuvenated by science and what can be done to avert the damage. In one story, a scientist, indeed, found a way to destroy Godzilla, but this would involve using a powerful weapon that would also destroy other living beings. In this story, the scientist decided to end his own life so that his secret would not be spread further. This is just a fiction, but it poses a question as to what a scientist in real life would do when confronted with such a dilemma.

## 9.2 Risks from Science and Technology: Assessment and Management

Before an output of science, be it new drugs, new gadgets, or new industrial processes, can be fully offered to society, it has to be fully tested and the risks of harm reduced to an acceptable level. Nevertheless, the output of science can sometimes lead to unexpected damage or harmful side effects. An example is the use of the drug thalidomide around the middle of the last century to lessen nausea and morning sickness in pregnant women. Unfortunately, the drug caused malformation of the limbs and other organs of the babies and even death. Another example is the widespread use of DDT as an insecticide, which damaged the environment in many ways, such as the reproduction of birds. These examples show that we can sometimes look for only the benefits of science, with too little regard for the possible unexpected harms.

If there is something good coming out from these examples, it is that we have learned to be more careful in adopting new products and processes introduced by science. Regulatory agencies in countries with major research and development activities now have strict requirements for safety tests for new drugs, vaccines, food additives, cosmetics, workplace chemicals, household products, and many other substances that may pose a risk to consumers. New drugs and vaccines have to go through preclinical testing in laboratories for various aspects of safety, including toxic and other effects on laboratory animals. Those that pass the tests have to undergo at least three phases of clinical trials with human volunteers to check for safety and efficacy. Countries around the world have environmental impact assessment processes for projects, programs, and plans that may have a negative impact on the environment. These include activities that would release agents into the environment, interfere with biological processes, or affect natural processes concerning energy, water, or the climate. Scientists as well as other people involved have to abide by regulatory procedures, which were developed through scientific principles coupled with lessons from past experiences. Science therefore plays an important role in correcting and preventing adverse effects of science itself. Scientists also tend to look for more gentle ways to achieve results in, say, agriculture or energy procurement rather than using the brute force of disruptive technology. For example, in protection of crops from pests, consideration is made for use of agents other than synthetic chemicals, such as biological agents or bioproducts that do not have long-lasting effects on the environment. Integrated pest management is also increasingly used, relying on natural control of pests by their predators and a mix of mechanical (e.g., traps, tillage, cover, and even human picking), biological, and chemical control processes. These can be used together with other means, including the use of pest-resistant varieties and the harvest and use of insect pests as food. These strategies have lower risks to the environment and to the development of sustainable agriculture, where people and nature can coexist and flourish together.

Like other activities with an impact on health and the environment, risks associated with the introduction of new agents or processes resulting from scientific development need to be thoroughly assessed. The assessment must concern both the likelihood of adverse effects and the potential damage that they may cause. This may be done through at least three steps: identifying the hazards involved, analyzing the effects of the hazards (e.g., through dose–response analysis), and estimating the amount of exposure to the hazards. The results of the three steps are combined to give an estimate of the total risks. The risks have to be weighed against the benefits that the undertaking is expected to bring about so as to come up with an optimal decision. The risk assessment process is not easy, even with familiar processes like introduction of a new cosmetic or building of a hydroelectric dam. For areas in science and technology, this is even more difficult since we are not familiar with the risks involved, and some hazards may be underestimated, while others may be overestimated. In some cases, the issues are complicated by conflicting risks. The use of DDT as an insecticide is a case in point. The work of Rachel Carson on the adverse effects of DDT on the environment led to a ban on its use in many countries. However, developing countries with major insect-borne diseases like malaria cannot afford to change to other, more expensive insecticides, and the World Health Organization (WHO) declared its support for continued indoor spraying of DDT in these countries.

Once the risks have been identified and their potential effects assessed, further action consists of prioritizing the risks and considering the range of options that can be taken. These include stopping the project, making further study, or going ahead with suitable prevention and mitigation measures. Prevention of risks can be achieved by choosing alternatives that do not carry the risks. In a project for synthesis of a drug, for example, some steps may carry the risk of explosive reactions, which can be prevented by choosing other reaction steps that do not carry such risks. Risk mitigation is a reduction of risks to manageable or acceptable levels. In the example of the chemical synthesis, if it is necessary to carry out potentially hazardous reactions, the reaction must be done in a controlled environment and steps taken to ensure the safety of the chemists and the facilities. Risk prevention and mitigation are important in all activities where people and the environment may be exposed to the effects of adverse events, such as mishaps in power plants, release of new pesticides, or release of excess heat such as that from industrial processes. Special precautions need to be taken to ensure the safety of people and the environment from new agents and procedures in science and technology, including communication to the public concerning the nature of the risks and measures taken to prevent or reduce the risks. The *precautionary principle*, often applied by policy makers, states that when an action or policy has a suspected risk of causing harm to the public or the environment, the burden of proof falls on those taking the action or adopting the policy. This principle should be applied to all projects carrying risks for the public, including those in science and technology, although in many cases it is difficult and contentious, requiring assessment of risks that may be unfamiliar or difficult to gauge.

Since no action has a true zero risk, an acceptable risk for a project could be set at, say, less than a one-in-ten-thousand chance of a lifetime risk. For new ventures in science and technology, such quantitative risk criteria may not be possible, and each risk may need to be studied qualitatively, together with prevention and mitigation measures. In some cases, a risk matrix may be constructed, in which one dimension is the probability of adverse events and another is the impact of damage (see Fig. 9.1). Where possible, quantitative figures (e.g., the number of people affected or the costs of damage) are estimated and definitions of acceptability given after comparison with potential benefits accrued. While this type of analysis may be possible with known activities such as building of a power plant, it is not adequate for a novel science and technology undertaking in which in-depth qualitative analysis and consideration of various effects are not easily quantifiable.

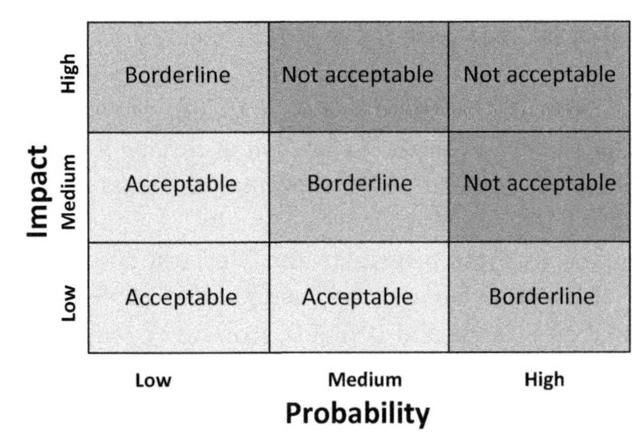


**Figure 9.1** An example of a risk matrix, where adverse effects of a project are analyzed in terms of probability and impact. The matrix can be enlarged to contain more boxes, as appropriate. Wherever possible, levels of acceptability for the boxes are defined in quantitative terms. Projects in the borderline areas can be modified so as to mitigate or reduce the risks.

## 9.3 Ethical, Legal, and Social Implications of Science and Technology

The fruits of science and technology can be used to bring many human benefits. However, science and technology can also be used to bring harm, and they can also have unexpected side effects that are harmful. Some achievements in science may be like a box of mysteries that should not be opened without adequate safeguards against things that may go wrong. There is a need to consider science as not only a technical subject but also a subject that has ethical, legal, and social implications (ELSIs). For example, the ability to edit genes of living beings, including humans, necessitates the consideration of various issues, including human rights and welfare, social justice, and the future of the living environment. The rapid development of robots, for both industrial and household use, calls for debates on the future of human jobs threatened by the robots; legal responsibilities of robot makers, sellers, and users; and the morality of wars conducted by remote control. Information technology (IT), which has become powerful and pervasive, threatens the preservation of privacy of persons and families and the integrity of confidential information.

These are some of the issues that will be considered in more detail next.

An important consideration is respect for human rights. According to the "Universal Declaration of Human Rights," adopted by the United Nations in 1948, important elements of these rights are principles of dignity, liberty, equality, and brotherhood. In accordance with these principles, the "Declaration on Science and the Use of Scientific Knowledge," adopted in 1999 by the World Conference on Science and UNESCO, stresses the urgency of using scientific knowledge in a responsible manner to address human needs and aspirations without misusing this knowledge. Through these and other professional ethics, scientific associations and the majority of scientists all over the world are aware of the importance of doing their work in a responsible manner and are willing to listen to the public when their work may have issues concerning ethics and society.

We should also consider science as an integral component of what makes us human. It is often said that the three enduring values of humanity are truth, beauty, and goodness. Many people associate truth with science, beauty with arts, and goodness with spirituality. While this may be accepted as a broad generalization, we should be aware that it is oversimplification, since science is, in fact, associated with all three of these enduring values. Science deals with finding possible truths, leaving room for correction of errors and improvement of purported truths. What science finds are things of beauty in the sense that it gives us a greater understanding of nature and also tools with which to create beauty such as modern works of arts and to preserve existing arts in forms that can be appreciated by more people, such as on the Internet. Truths are also ingredients of goodness, since from truths we can go on to build moral codes and yardsticks for justice. Legal judgments, for example, rely increasingly on evidence provided by science, including deoxyribonucleic acid (DNA) and medical records, pictures, films, and sound recordings.

## 9.4 Bioethical Issues

*Frankenstein*, a story written by Mary Shelley in 1818 and reputed to be the first science fiction, tells the story of a science student who created a gigantic and hideous living creature from lifeless things. Both the monster and he were entangled in various tragedies resulting from this action. Today, through synthetic biology, we are able to create life from genes assembled from chemical processes. We hope that our creatures will yield useful products like drugs and vaccines, but will our lives become entangled in unexpected events not to our liking? Will we inadvertently create our own monsters out of our control? While the Frankenstein story is fiction, in real life the science of the living system has now progressed to the stage when we should think of its ethical and safety implications similarly to the fictional situation. Science now has the ability to transplant tissues and organs from other people, and possibly soon from nonhuman species. We can anticipate the day soon when tissues and organs can be repaired or regenerated from stem cells from our own body or someone else's, perhaps on scaffolds made from 3D printing. We can read the genomes of living species, including ours, with great speed and can edit them. Genome editing can correct the mistakes residing in the genes of some people with genetic diseases, but it can also potentially change the genetic makeup of plants, animals, and people and their offspring. We have drugs and surgical methods that can alter brain and neural functions. All these advances are mostly intended to be used for good causes like curing people from sickness, but how can we prevent them from being instruments of unscrupulous people or causing unintentional damage?

These issues need understanding and reflection of their implications by the scientists together with policy makers, philosophers, theologians, lawyers, journalists, and the public in general. First, what is technically feasible may not be ethically acceptable or may be acceptable in some cases and not in others. For example, cloning of human beings may be technically feasible in the future, but although this is favorable to some, others have serious ethical objections on moral, religious, and other grounds such as human dignity. Transplanting of organs and stem cells has ethical issues such as commercialization of the source, which is illegal in most countries but is still illicitly practiced in many. In some countries with weak laws or law enforcement, the organs may be from unwilling or unknowing sources, such as condemned prisoners. The use of embryos as sources of stem cells for transplanting is technically possible and promising as therapy for many diseases but is ethically contentious, and a continuing debate is needed as science progresses.

Knowledge about our own genes, some of which may be defective, is desirable to some but an anathema to others who may not wish to know. Who has the right to this knowledge, apart from the person who owns the genes? The parents, the offspring, health authorities, or insurance companies? When it is technically possible to edit our own genes, which is a near possibility, under what circumstances should people be able to do so? And should we have the right to edit genes of our offspring and future generations? These are just some of the questions that need to be tackled by the nonscientific parties as well as the scientists, through reflection, debates, and consideration from various angles.

Some areas involving risks to the public or individuals and ethical considerations concerning health, agriculture, and the environment from science or its products are clear-cut. For example, antibiotics are often used as additives in livestock feed, with the result that the animals grow better and have fewer infections. However, residual antibiotics in animal meat give rise to microbial resistance against these drugs. This problem has been recognized in many countries, and such practice is now unlawful. Other issues are more contentious, such as surrogate motherhood, or the practice of carrying a pregnancy and delivering another person's baby. Countries and societies vary with respect to legality, ethical, and cultural considerations. Test-tube babies from in vitro fertilization have been realized for more than a generation now, but there are ethical concerns about the possibility of choosing the sex of the babies. Would this involve destruction of female embryos? And how would we face a world where men will outnumber women by a large margin, enabled by such technical possibility? Genetically modified organisms (GMOs), such as plant varieties with extra genes inserted for higher productivity or resistance to diseases, have also long been available, but many countries still do not allow the introduction of such organisms, ostensibly due to fear for consumers' safety and environmental hazards.

Yet there are other issues where science has gone even further than what we are used to dealing with in ethics. For example, we can anticipate that gene editing will soon be applicable to microbes, plants, and animals, including humans (see Chapter 7, Box 7.2). How would this be regulated so that we can produce benefits to people and the environment and not inadvertently harm the planet and our society from unrealized consequences? These issues require deliberations and reflections among the scientists and the public, including ethicists, sociologists, lawyers, journalists, and others concerned. The scientists need to lay the groundwork for debates among the various people concerned by first talking among themselves to clarify the state of issues and anticipated possibilities. We faced such a dilemma in 1975, when it just became possible to manipulate genes through DNA technology. The Asilomar Conference on Recombinant DNA was organized in California, with participation of biologists, physicians, and lawyers, to consider the implications of the technology and draw up voluntary guidelines to ensure the safety of the technology according to the precautionary principle. Among the outcomes of the conference were the principle and practice of containment of potentially dangerous organisms and materials through physical and biological barriers and the prohibition of experiments on, or production of, highly toxic bioagents or organisms. Forty years later, another summit meeting was convened, attended by scientists, ethicists, sociologists, legal experts, and advocacy groups. The Summit on Gene Editing, held in December 2015 in Washington DC, was devoted mainly to new issues resulting from our ability to edit genes of embryos and after birth. While acknowledging the great potentials of being able to correct genetic defects, the meeting was concerned that the possibility of gene editing in embryos can result in changes not only in the offspring but also through succeeding generations. Many also worry that the technology could create inequality and discrimination in the future among those who can and those who cannot afford the technology. The meeting came out with a recommendation for the researchers to not stop human-gene-editing research outright but to proceed cautiously and be aware of risks such as harmful effects from inaccurate or incomplete gene editing and effects on the human population and evolution. The meeting also agreed on the need for an ongoing forum among all the stakeholders to try to establish norms concerning acceptable use of human gene editing.

Concerns with ethical, societal, legal, and related issues with life sciences and biotechnology have led many countries to have laws, regulations, and guidelines on these issues, both at national and at institutional levels. The subjects of such concern include human and animal experimentation and the environmental impact of released products or living systems. At the international level, UNESCO and other nongovernment agencies have arranged programs and activities to alert and give guidelines to scientists, institutions, and the society in general on outstanding issues in bioethics. For example, the UNESCO Bioethics Programme has a continuing mandate on bioethics and activities to strengthen the capability of developing countries to anticipate and deal with these issues.

## 9.5 Ethical and Related Issues on Information Technology and Robotics

The age of IT that we have entered since around the end of the past century has brought many benefits and promises to bring substantially more. Vast amounts of data on health, financial, and personal information can be collected and analyzed for various purposes, ranging from disease risks to banking and market strategies. However, together with the benefits come risks and many ethical, societal, and legal issues, some of which are quite new, while others are familiar but present new angles. Examples of the familiar issues are those concerning protection of privacy and confidentiality of information. IT has made it possible to store and transmit vast amounts of information, substantial parts of which concern sensitive or confidential financial, trade, security, and personal issues. In the old days, such information was stored as hard copies, mostly paper, but IT has made it possible to store it as electronic files and other forms. While confidentiality can be ensured with encryption technology, electronic codes, and security passwords, unauthorized access can be gained by theft of encryption keys, security codes, and passwords or by other illicit means. E-mails and other means of information transmission through the Internet can be spied on without knowledge of the owners of the information. The stolen information can be used in many unethical and illegal undertakings, including fraud, pornography, and plagiarism, to name a few.

Computers and the Internet have made it possible not only to steal information but also to disrupt information and communication through various means. They are increasingly used in what is called the "Internet of Things," which uses sensors and controls to monitor and run various devices such as gadgets, houses, and cars. Although the controls have security settings, these can be overtaken and used maliciously by ill-meaning people. Computer viruses and worms are software programs that can infect and disrupt other computer programs by modifying them. Hacking is the act of breaking into computer systems and accessing and stealing the data and programs without knowledge of the owners. The use of these new means of illegal methods of attacking opponents or for personal gain is called cybercrime. These are challenges to developers of computers and communication systems, who must safeguard systems against such illicit attacks. In addition to physical security, lawmakers must now also be technologically prepared so as to be able to maintain what is called cybersecurity. At the international level, the issue of cybersecurity has now assumed importance comparable to physical security. Many countries, including the United States, the United Kingdom, South Korea, and Estonia, have been subjected to coordinated attacks, which caused disruptions of a number of public and government services. Some issues are ethically complex, with moral judgments that can be considered both wrong and right. Consider the example of WikiLeaks, which is a whistle-blowing website dedicated to expose misconduct by governments and corporations. While this is hailed by many as good investigative journalism, it is condemned by others as morally wrong in exposing secrets that may endanger the lives of innocent people. These issues illustrate the moral dilemma in dealing with secret information, a long-standing problem that has been brought to the fore because of the ease of transfer of vast amounts of information made possible by technology.

Good education can help us in dealing with ethical, legal, and social issues stemming from the use of IT. Technical education alone, building capability in using and developing computers, cell phones, and communication media, is not enough. We also need education to make us responsible and ethical members of society, both in general and with respect to issues concerning IT in particular. We need to respect the privacy of individuals and groups, observe the legality of actions, and have sensitivity to issues such as gender, age, disability, race, and social status of people. These issues are not new but are given heightened priorities with the immense power of information and communication technologies to amplify the effect of individual action, such as the use of social media.

New laws, called IT laws, are needed to govern the access and use of digital information and software and other cases of use and development of the technology. Laws concerning the Internet, sometimes called cyberlaws, are also needed either as new laws or as amendments to already existing laws on intellectual properties, commerce, and freedom of expression. As in other areas, the laws need to be fine-tuned so as to address criminal aspects without suppressing political freedom or infringing too much on individual liberty. Guidelines from international and national bodies are helpful in dealing with problems arising from information and communication technologies. In many cases, individual judgment is important, with due reflections on various aspects, with morality as the overriding principle.

More ethically complex issues emerge when we deal with artificial intelligence. This has its roots in both IT and other sciences dealing with consciousness, perception, and brain-based decisions and actions. It covers areas ranging from optical character recognition and language processing to search-and-rescue tasks and self-driving cars. Artificial intelligence incorporated in mechanical devices gives rise to robots. Robots are smart machines that perform tasks under human command or on their own with embedded computers and software programs. In addition to industrial applications, robots can or will soon be able to do many tasks that require intelligence, including being learning tools, giving domestic help, aiding sick or old people, performing surgery, and being smart toys. They can be used in surveillance and policing, helping in search-and-rescue work, or act as autonomous weapons. They can be so small and agile as to act as spies that are difficult to detect. Although there are many potential uses of artificial intelligence and robots, which will soon play a big part in our everyday lives, eminent people in science and technology, like Stephen Hawking, Elon Musk, and Bill Gates, have warned of the danger that they are evolving faster than the human race and may one day spell doom for humanity if we are not careful in assignment and control of their uses.

The word "robot" was first used in a Czech play and comes from the word "robota," meaning work or labor. Indeed, intelligent machines have captured the imagination of countless generations, as seen in examples ranging from Greek mythology to the robot cat Doraemon in the popular Japanese series. Although in popular notions, robots are assumed to be humanoid, looking like and functioning as humans, they can take various forms and functions, ranging from drones to industrial manipulators. The essential features that require ethical consideration are interactivity with humans and the environment and potential autonomy. Other issues include perception, searching, learning, reasoning, planning and problem solving, and storing and use of knowledge, all of which result in decisions and actions that have ethical implications. The use of robots in various tasks brings up the issue of the future of work for humans. Many people are worried that robots with high intelligence and manipulative dexterity may put people out of their jobs. The use of robots can also bring up complicated legal issues, for example, the identification of liable parties when an intelligent robot is involved in an accident or a premeditated harmful event. Who is responsible when a self-driving car has an accident or when two-self driving cars crash into one another? The issues will become more complex when robots have more intelligence and autonomy. Would the incident in which the US police used a robot to kill a suspect in a shooting incident be a pointer to deployment of real, not movie-version, RoboCops of the future? What about robot security guards employed by citizens? What about robots used by criminals?

The science fiction writer Isaac Asimov foresaw these problems and invoked the three laws of robotics in a short story written in 1942:

1. A robot may not injure a human being or, through inaction, allow a human being to come to harm.
2. A robot must obey the orders given it by human beings, except where such orders would conflict with the first law.
3. A robot must protect its own existence as long as such protection does not conflict with the first or second law.

Another law, the zeroth law was added later:

0. A robot may not harm humanity or, by inaction, allow humanity to come to harm.

Although these laws can form the basis of real laws and regulations concerning robots and artificial intelligence in the future, they can serve only as general guidelines and they require more specific considerations relevant to the present world, not to a world far in the future. Moreover, laws and regulations would be just one tool and by themselves inadequate to cope with the implications of artificial intelligence. Codes of ethics for making and interacting with robots have been and are being developed by professional and industrial organizations. Since robotics is a new subject, these codes will need to be constantly revised and updated to include new aspects of human–robot interactions. Important principles for formulating these codes include consideration of human safety, dignity, and privacy, as well as the soft side of human nature, such as emotions and the liability to make errors. As robots (Fig. 9.2) become more and more like humans, scenarios of humans falling in love or engaged in other emotions with robots will become more common. In the fiction *Bicentennial Man*, Andrew, an intelligent robot, becomes more and more like humans and finally asks to be operated on so that he can be a full human and become a true partner of the woman he loves. Should robots be designed to look and act more like humans or more like machines? A humanoid look of robots makes it likely to induce attachment by human users, and robot designers and manufacturers should consider the emotional consequences of human–robot interactions as important issues. This and other delicate issues concerning the ethics of human–robot interactions will become more important as robots acquire more intelligence and autonomy in the near future.

**Figure 9.2** "Now, what is the third law of robotics?"

## 9.6 Environmental Ethics

Human beings have changed the environment to a far greater extent than any other living species. There is increasing concern that we may change it so much as to exceed the finite planetary boundaries with regard to pollution, climate change, water use, extraction and use of fossil fuels, and destruction of biodiversity and other areas (see Chapter 2, Section 2.6). It is clear that we cannot go on in the business-as-usual manner, and environmental ethics needs to be taken into account in order to build a sustainable world. Environmental ethics is concerned with the rights and wrongs of things that affect not only humans but also the environment as a whole, of which humans form a part. We have a moral obligation to conserve the environment and ensure its sustainability. This obligation is not only confined to the immediate environment around us but also extends across national boundaries to cover the whole planet, and indeed the space beyond. Moreover, as it is generally agreed that we are now in a new geologic epoch, called the Anthropocene epoch, in which human activities have a great impact on the planet, we should be concerned not only with the present but also with the future of the environment with increasing human influence.

Environmental ethics stipulates that humans should not only be concerned with what is good for humans, or anthropocentric issues, but also be concerned for the environment as well. As Aldo Leopold, an early ecologist, said, "A thing is right when it tends to preserve the integrity, stability, and beauty of the biotic community. It is wrong when it tends otherwise." We can go further to include the whole planet, not just the biotic community. Many ethical concerns about the environment have consequences for humans as well, for example, concerns about the extravagant use of mineral resources, which will impact our need for them in the future. In addition to general concerns about preserving the status of our planet both at present and for the future, environmental ethics considers issues about the environment in relation to rights, justice, fairness, compassion, and other moral aspects. Human rights include the rights to good living conditions, food, water, and health, all of which depend on a healthy and sustainable environment. In addition to human rights, we might also consider the rights of animals and biodiverse communities. Actions that lead to the extinction of a species, for example, can be considered as similar to crimes committed on humans. As for environmental justice, the topic can be considered from the point of view of justice to the human society and justice to the environment as a whole. As an example, actions that cause global warming, such as the generation of greenhouse gases by a country, can be considered unjust as global warming affects the whole global community, not just the generator of greenhouse gases. Moreover, global warming is a long-term process that affects the future as well as the present, causing injustice to the future planet and our own future generations. Environmental ethics is also concerned with compassion for the failings and sufferings of others, which include both humans and the whole planet of the present and the future. These principles should make us consider the state of the environment in various aspects and prompt us to take actions to avoid harm, mitigate damage, and contribute positively to the state of the environment.

Science has a major role to play in environmental ethics, both in defining the big picture and in analyzing the benefits, risks, and ethical concerns of various areas of human activities. Ecology is a branch of science that deals with relations of organisms with the environment and with one another. It is an ethical concern to maintain the health of an ecosystem or a community of living organisms and their surroundings. This has to be balanced with benefits to humans in using the land, water, forests, or other parts of the ecosystem for their own purpose. In an even bigger picture, environmental ethics has to play a part in the ethics of sustainability, which takes into account not only environmental aspects but also economic and societal aspects, such as consideration of freedom from hunger, supply and use of energy, and decent work and economic growth. Each specific area can be considered in detail with the aid of scientific analysis. For example, extensive agriculture can help solve the problem of hunger, but it must come at the cost of deforestation, high water requirement, and potential water pollution from excess fertilizers. Nuclear energy can provide a large supply of energy, but it carries a risk in terms of potential radioactive contamination of the environment. Work and economic growth can be obtained through industrialization, but this will generate greenhouse gases, which cause global warming. These considerations can be made with the help of a risk matrix that was introduced earlier (Fig. 9.1) and other analytical tools. The risk matrix, an array of risks assessed in terms of probability and severity, gives a rough guide to decision makers. It must be pointed out that scientific analysis can give only information and conclusions based on data, but the eventual decisions on the courses of action depend on ethical judgments that take into account all the factors mentioned before, including the sense of justice, fairness, rights, and compassion both for the human society and for the planet as a whole.

## 9.7 Toward Good Governance in Science and Technology

Scientists should have freedom to do research and explore possibilities of the use of research findings. However, scientists should try to make sure as much as they can that their work does not lead to harm to individuals, the public, or the environment. With recognition that science can go wrong, we should learn from past lessons and try to avoid mistakes from science in the future. This effort is all the more important, considering the fact that science plays an increasing role in our lives, both in everyday affairs and in matters concerning health, trade, industry, the environment, and international relations. These are matters for concern, not only for the scientists, but also for the people in various professions and the lay public. Just as in other areas of activities that affect the public, we need good governance in science, technology, and related issues. Such good governance requires observance and participation by all the stakeholders, the roles of whom we will now discuss briefly.

We first consider the role of the scientists. With public trust in scientific knowledge, it is of utmost importance that scientists acquire such knowledge with great effort and interpret it for the public with integrity and honesty. They should clearly point out the implications of their findings with regard to public interest, especially if the research is funded with public money. If it is later found out that the work that they reported earlier is erroneous, they have to retract it with a due apology. While a great majority of scientists observe this principle, there are some unscrupulous scientists who intentionally fabricate their work for their own personal gain or other motives. Fortunately, science is an activity that is reproducible, and sooner or later the fabrications will be found out by other scientists who want to repeat the work, especially if the work is important to the field. A few cases of fake science findings are illustrated in Box 9.1. Data fabrication and false reporting are considered a serious misconduct in science. Also serious is misconduct arising from knowingly doing experiments that endanger the lives of people or the environment without good justification. This includes not just the experiments themselves but also the consequences that arise from their conclusions, for example, drugs with adverse effects or insecticides that harm the environment. This is why scientists must seriously investigate potential harms or risks of their initial discoveries, through controlled toxicology or environmental assessment, so as to ensure their safety.

**Box 9.1** Scientific Misconduct: False Claims of Scientific Discoveries

The progress of science depends on scientific discoveries made by scientific researchers. The career of a scientist, therefore, to a great extent depends on the discoveries that he or she makes and reports to the world in scientific publications. Scientists, by and large, respect the cardinal rule of honesty in reporting the results of their investigations. However, through a desire to succeed professionally or other motives, some scientists falsify or fabricate the works that they report to the scientific world and the public. Others steal the work of others and claim the discoveries to be their own. Yet others just want to create hoaxes for their own satisfaction.

One of the most celebrated cases of hoax was the Piltdown Man. In 1912, an amateur archaeologist, Charles Dawson, claimed to have found fragments of a skull and jawbone from a gravel pit at Piltdown, Sussex. He and another investigator claimed them to be the fossilized remains of an unknown early man with the characteristics of an ape. The discovery was exposed in 1953 as a forgery, a construction consisting of the lower jawbone of an ape combined with the skull of a fully developed, modern man. The person who committed the forgery remains unknown but is widely thought to be Dawson himself.

In 1998, Andrew Wakefield et al. published a study that purported to show that there was a link between administration of the vaccine against measles, mumps, and rubella (MMR) in children and autism and bowel disease. This caused a great scare, and after the publication of this study, many parents refused to let their children be immunized. Subsequent investigation by other researchers failed to support his allegation, and serious flaws were found in Wakefield's study. It was also revealed that he had filed a patent for a competing vaccine, a case of conflict of interest. The journals in which he published the work and related findings retracted the paper, as did most of his coauthors. In 2010, the UK General Medical Council found him guilty of misconduct and erased his name from the medical register. The harm had been done, however, as unvaccinated children were exposed to greater risk of the diseases, and antivaccine activists continue to use the discredited work in their campaign against vaccination in general.

The promise of stem cells in therapy against various ailments, including the repair of aging, has prompted furiously competitive research. The quest for procedures to make and use stem cells has brought fame to many who succeeded but also downfall to those who resorted to fraud. A related technology of cloning can produce genetically identical animals of value as pets or in farms. Around 2004, Hwang Woo-Suk, a South Korean researcher who had already made his name from his ability to clone cows and dogs, claimed that he could clone human embryos and produce cloned human stem cells. It was later revealed that he obtained the human eggs for the experiment in an unethical manner from his graduate students and paid donors. Worse still, further investigation by the Seoul National University, where he worked, found that he fabricated the results of his experiments, and the university subsequently dismissed him. He was later tried and convicted of embezzlement and bioethical violations.

A more recent case of fraud concerning stem cells emerged in 2014, when Haruko Obokata claimed that she could trigger normal cells into an embryonic-like stage simply by subjecting them to stress such as pressure or acid. Investigation by Riken, the institute where she worked, showed that she manipulated the data and lacked ethics, integrity, and humility as a researcher. The papers were retracted, and Obokata resigned from the institute. One of her coworkers committed suicide, and Riken's president resigned over this incident.

Scientists must be careful to avoid conflicts of interest in their work. For example, in research funded by private companies, there may be pressure to report findings that favor products made by the companies, which the scientists must resist. In any case, they have to declare honestly that their reported work is free from conflicts of interest. Scientists must also be mindful about possible misinterpretation of their research findings and explain the results and their implications to the public so as to prevent such misinterpretation. An agent found to prolong the life of, say, mice should not be grandly introduced to the public as an elixir of life. Scientists should also recognize any possibility that their work may lead to harmful or risky consequences, such as being used for chemical or biological weapons, and avoid reporting to the public in such a way as to assist wrong doings. In doing their work, scientists should respect the dignity and rights of people who may be involved as research subjects or in other ways.

Many scientific professional organizations have issued voluntary codes or guidelines of conduct on matters involving ethics and risks in science and technology. These codes and guidelines (Box 9.2) are akin to the Hippocratic Oath in the medical profession. They deal with establishing principles for best practices of scientists in general or for specific issues such as manipulation of genes or the development of nuclear power. UNESCO has issued the "Universal Declaration on Bioethics and Human Rights," which serves as a guideline for ethical conduct in bioscience and biomedical research. Civil societies and individuals can also help warn the scientists and involved parties about the risks and potential pitfalls of science. Many scientists have been active in warning the public and governments about risks in the misuse of the products of science. The Pugwash Conferences on Science and World Affairs were launched in 1955 with a manifesto initiated by Einstein and Bertrand Russell, warning the world of the danger of nuclear weapons. For many decades now, scientists have warned humanity about looming environmental crises resulting from unsustainable ways of life that we have led, with global warming, loss of biodiversity, and threats to water and land resources as consequences. These messages to the public also serve to warn the scientists themselves to avoid being engaged in activities that may bring harm and unacceptable risks to humanity and the environment.

**Box 9.2** Codes of Conduct for Scientists

Doctors have to adhere to good principles of practice because of their responsibility toward patients and other people affected by their actions. The medical profession has codes of conduct, to which doctors must adhere and a violation of which can lead to serious outcomes, including a ban from practice. The Hippocratic Oath, an ancient oath ascribed to Hippocrates 25 centuries ago, is still taken by many medical schools today. This and modern forms of codes of conduct, such as the one called the "Declaration of Geneva," require doctors to strive for the health of the sick and avoid harm and mischief. Because of the potential of science to cause harm, some concerned scientists have advocated the equivalent of the Hippocratic Oath for the scientific profession. The idea was suggested by Sir Joseph Rotblat, a Nobel laureate, who was a chief advocate of nuclear disarmament. The idea has not been taken up universally by the scientific profession, perhaps because of the hesitation of many scientists to be bound by an oath. However, many professional organizations and some governments have issued codes of conduct to ensure ethical practice in science. The UK government, for example, has adopted a universal code of ethics [45], with the following seven principles:

- Act with skill and care in all scientific work. Maintain up-to-date skills and assist their development in others.
- Take steps to prevent corrupt practices and professional misconduct. Declare conflicts of interest.
- Be alert to the ways in which research derives from and affects the work of other people, and respect the rights and reputations of others.
- Ensure that your work is lawful and justified.
- Minimize and justify any adverse effect your work may have on people, animals, and the natural environment.
- Seek to discuss the issues that science raises for society. Listen to the aspirations and concerns of others.
- Do not knowingly mislead, or allow others to be misled, about scientific matters. Present and review scientific evidence, theory, or interpretation honestly and accurately.

Research and education institutes, including universities, can play big potential roles in developing good governance for science and technology. In addition to expanding the technical knowledge in scientific areas, they should also cover the subject of ELSIs of science and technology. Knowledge and understanding of such issues come from integration with broad social and policy studies. Many universities now have programs in ethics of science and technology and related subjects. Universities can also help generate healthy debates among scientists, various stakeholders, and the public on issues that impact society, such as robotics, gene editing, and organ transplant.

Governments have a role to play in overseeing development and capability strengthening in social and ethical issues in science and technology. In some cases, such as the use of IT, they can set up regulations and laws, as necessary, making sure of a healthy balance between public interest and scientific advance. For example, privacy and human rights need to be respected in the age of gathering and using big data. Governments need to understand, and can help make the public understand, scientific issues that have ethical and related implications. Dialogues are needed among the parties involved, including experts, decision makers, civil society, and the lay public. Many government agencies, including research and public service agencies, have rules and codes of conduct for their professionals that are based on the results of such dialogues and other considerations.

## Chapter 10

# Sparks for Sustainable Development

*Sustainable development is development that meets the needs of the present without compromising the ability of future generations to meet their own needs.*

—*Our Common Future*, Report of the UN World Commission on Environment and Development (Brundtland Commission), 1987

In search for sustainability, we can learn from nature and past societies about why ecosystems and societies collapse and what make others sustainable. Threats to sustainability of human societies include wars, emerging diseases, poverty, economic instability, water, energy, uncontrolled population growth, and climate change. Problems with sustainability of our present society and the environment were realized some decades ago, but global attempts to solve them have been slow and have met with many obstacles. Presently, with support from all nations in the UN system, the problems are being tackled, with 17 goals for sustainable development set to be achieved by 2030. Science has a big role to play in the efforts to meet all the 17 goals of sustainable development. The role of science must be integrated with the roles of other areas of activities so as to achieve a balanced approach to sustainable development.

*Sparks from the Spirit: From Science to Innovation, Development, and Sustainability*
Yongyuth Yuthavong

ISBN 978-981-4774-57-4 (Hardcover), 978-1-315-14599-0 (eBook)
www.panstanford.com

## 10.1 Lessons from the Past

A major theme of this book is that science and its allied technology are major factors leading to innovation and development in various areas of human activity. The fruits of science have led to positive outcomes in the way we live, our health, and wealth in general. The picture is not all rosy, since the benefits have not been evenly spread, and many risks and dangers are associated with the applications of science and technology. Science has been used as an instrument of wars and crimes and has created unintended effects such as industrial pollution. Taking all things together, in spite of the drawbacks noted, science should not only be aimed at short-term development but also comprise important tools for sustainable development. Sustainability of the human society and the ecosystem means their continuity into the distant future in healthy states. Sustainable development can be defined in terms of development, given by the Brundtland Commission, as quoted at the beginning of this chapter. Better still, sustainable development should go further than meeting "the needs of the present without compromising the ability of future generations to meet their own needs" to also actively support and lay down the foundation for the ability of future generations to meet their own needs. Science can play a big role in realizing such an outcome.

In 2015, the United Nations adopted 17 goals for sustainable development, and we will examine the crucial role of science toward their achievement. Before doing so, we should learn some lessons from the past on sustainability and the collapse of ecosystems and civilizations so that we can note the important features of unsustainability. Many collapses in the past were due to natural circumstances, which we can do little about, but others were due to human mistakes, which we should try to avoid in the future. While collapsed societies offer lessons on mistakes to be avoided in the future, we can learn how some societies managed to avoid collapse and even flourished after resolving the crises of sustainability.

### 10.1.1 Ecosystems

We first examine the sustainability and collapse of ecosystems. Ecosystems are systems of diverse organisms living together in nature. They have natural resilience and can spring back from natural calamities or those caused by humans. In sustainable or stable ecosystems, the diversity of organisms is maintained through interactions among themselves and the environment. In such ecosystems, energy from the sun is trapped by green plants and some microorganisms and converted to chemical energy in the form of food and organic matters. The plants, which can be called producers, or their products, such as fruits, are eaten by primary consumers, squirrels and deer, for example. These may, in turn, be eaten by secondary consumers, such as birds of prey, tigers, and humans. When the consumers die, organic matters are returned to the ecosystems through decay and as feed to scavengers. A sustainable system is achieved when there is a steady energy flow through the system, with participation of various members through food chains and food webs. There may be some stress to the system from time to time, such as droughts, floods, or storms, but a sturdy ecosystem has resilience and can spring back to equilibrium after a time.

Strong or extended disturbances to ecosystems can, however, bring about permanent collapse. Members of the community die out or move away from the area, which may be denuded or populated by only a few species. These strong disturbances may be due to natural events, such as volcanic eruptions, asteroid impacts, or climate change. Ecological collapses can happen in the seas as well as on land. One present ecosystem under threat of collapse is the Great Barrier Reef off the coast of Australia, in which coral reefs forming a natural habitat for fish and marine life are under the threat of destruction through climate change and other factors. In the past 500 million years, there have been at least five major mass extinction events [46]. The largest extinction event, around 250 million years ago, killed more than 90% of all species. The extinction of dinosaurs around 66 million years ago was probably due to a combination of events, including volcanic eruptions, capped by the impact of a large asteroid in the Gulf of Mexico, which caused a lingering climate change. Large and extended collapses can lead to the extinction of species and eventually can allow new species to arise or become dominant. Seen through geologic periods of millions of years, extinction events can give rise to the evolution of new species as well as the loss of old species. The loss of old species or the collective loss of biodiversity puts the life and health of the remaining species at risk since the web of life is disturbed. As members of a present-day species, humans are under threat by their own actions that result in the loss of other species on which we depend, directly or indirectly, for our livelihood and well-being. Many ecological collapses in recent times have been due to human deforestation, which, in turn, has devastating effects on human societies themselves.

### 10.1.2 Human Societies

Like ecosystems, societies or civilizations can develop, become stable, or collapse. The rise and fall of societies and civilizations depend on both natural and human factors. Natural factors include climate change, change in geography, and drought. Human factors include warfare and activities that lead to deforestation, soil erosion, and imported diseases. In many cases, some factors cause them to falter, only to receive death blows by some other factors. Jared Diamond [47] listed five main sets of factors for societal collapse: environmental damage, climate change, hostile neighbors, overdependence on friendly trade partners, and society's response to its problems. The fifth set is crucial in highlighting the role of humans in deciding their own fate.

Easter Island is a small, remote Pacific island, which used to host a strong Polynesian culture named Rapa Nui around a thousand years ago. The island is distinguished by the presence of hundreds of huge stone statues of humans, dotted mostly around the shoreline (Fig. 10.1a). Now mostly a barren terrain, archeological evidence shows that the island used to have palm and other trees and was subjected to intensive agriculture. The stone statues and the barren terrain bear witness to the collapse of the Rapa Nui culture. The most likely, main cause of the collapse is deforestation and soil erosion caused by people who lived in about a dozen territories. The huge sizes of the statues could be due to one-upmanship in competition among the territories. The construction and transport of the huge stone statues, called *moai*, and the platforms on which they were erected must have required enormous resources of food and wood. Another factor could have been the loss of palm seed shells due to rats, leading to the loss of palm trees. Eventually, the population dwindled and resorted to cannibalism. The people also suffered diseases brought by the European visitors who arrived in the early 18th century. Altogether, the decline of the Easter Island community offers a grim example of what can happen to a society that overexploits its resources.

(c)

**Figure 10.1** (a) Easter Island statues, known as moai (from Wikimedia Commons; author Aurbina). (b) Tikal, a Mayan city, in present-day Guatemala (from Wikimedia Commons; author Shark). (c) Angkor Wat temple complex, in present-day Cambodia.

Although there are mysteries still remaining about the collapse of the Easter Island society in the past, it is a relatively simple case of societal disasters. The collapse of other societies and civilizations offers more complex examples of unsustainability. The Mayan civilization (Fig. 10.1b), located in the Yucatan Peninsula of present-day Mexico and other Central American countries, offers an example of a more complex and prolonged series of collapses. Various cities and communities rose and fell over a period of some 4000 years. The Classical period, which began around the year 250, saw the rule by "divine" kings believed to be mediators between humans and the supernatural. However, the civilization was already in decline by the time of the discovery of the "New World" by the Europeans. A number of factors contributed to the decline and fall of the Mayan civilization. The Mayans relied on corn as their main source of food, with no cereal grains, and only a few large domestic animals. This limited agriculture could hardly provide enough food for expanding societies, especially with population increase and stratification into various nonfarmer classes, including soldiers and bureaucrats, who do not produce food. People were therefore forced to move into hilly areas, where they practiced agriculture, which turned out to be unsustainable due to soil erosion. Another main factor that contributed to the downfall of the Mayans was intermittent warfare among various territories, which were not united, partly because of the mountainous nature of the terrain, and were driven to conflict by the scarce food supply. Large areas could not be cultivated because they became no-man's land. Climate change with intermittent severe droughts eventually forced the final collapse.

Our third example of the loss of societal sustainability is the Khmer civilization, which reached its height of prominence from the 9th to the 15th century. With the capitals located at Angkor Wat (Fig. 10.1c) and Angkor Thom, in present-day Cambodia, the Khmer empire ruled over most of mainland Southeast Asia and during its peak held the largest preindustrial center of the world. It was ruled by a succession of kings with immense power and wealth. Particularly noteworthy was the advanced system of water management, with a network of canals and reservoirs known as barays. The advanced art of water management was vital to the health of the empire, which had its vulnerable points in having to cope with droughts and floods frequent in that tropical terrain. Indeed, it is believed that the decline and fall of the empire were due to prolonged droughts from climate change with which the water management system could not cope. This was followed by death blows from invasion by the Thai and other foreign armies. Disease and famine could also have been contributing factors to the collapse.

These examples, and many more, serve to show that societies and civilizations of the past could succumb to problems undermining their sustainability. They serve as a warning to our present civilization, which is also vulnerable to the same or similar problems, plus many more that we have created ourselves in modern times, such as problems from unintended effects of technology. Recurrent themes can be noted in these examples, including climate change, food supply, population, and wars. However, we should realize that many societies and local civilizations of the present day have survived through various crises, and we should also learn from them. For example, Diamond [47] pointed out that together with the collapse of the Easter Island society, there is continuity of the tiny society in Tikopia, an even smaller island in Polynesia. Key differences between the two cases are successful forest management and population control in Tikopia by the community itself. We can also note that while deforestation was a main factor in the collapse of the Mayan civilization, and is still today going on at alarming rates in many areas such as South America, Africa, and Asia, many countries have good management systems, leading to sustainable forests. Water management, a key factor in the rise and fall of the Khmer civilization, is still a major problem in many present societies. Hunger, starvation, diseases, and other areas affecting sustainability are still problems today as in the past. Wars, which were a major factor in the collapse of many civilizations in the past, are still occurring in many parts of the world today.

A main difference of present-day societies from past ones is the relatively much more advanced state of science and technology today. As we have seen throughout this book, they have been major tools in solving many problems of day-to-day living, agriculture, health, commerce, industry, and services. For example, they have enhanced connectivity of people worldwide, both physically in transport and through telecommunication. Crises that happen in one place, such as earthquakes and volcanic eruptions, can be mitigated through assistance from other parts of the world. Therefore, we face the problems of future sustainability together as one connected community, and this should make a big difference. Hopefully, science and technology can be major factors in our path to sustainable development, which we will now explore.

## 10.2 Roads to Realization of Our Problems

Our present civilization is in much better shape than past civilizations on many counts. Our average life expectancy is now 70 years and increasing, while only a century ago, it was less than half this number. Although hunger is still a major problem in many parts of the world, our total food production ensures that there will be no global starvation in the foreseeable future. Global industries and commerce provide goods, services, and connectivity to people worldwide. Although we are subject to periodic economic downturns affecting widespread regions or even the whole world, they have so far proved to be limited or temporary. This does not mean that we can become complacent and take a business-as-usual attitude. On the contrary, there are many disturbing signs that the global civilization is in an unhealthy state in many ways, and urgent actions are needed to stave off future crises. In many ways, we are like frogs in slowly boiling water, hardly aware of the rising heat. We need to jump out before we are cooked.

The disturbing signs came from various sources, at first dismissed by the business-as-usual crowds as alarmist and on the fringe. Over the past half century, the signs have become increasingly clearer and no longer possible to ignore. Among the first cries of warning was that from Rachel Carson, whose book *Silent Spring* [48] pointed to the detrimental effect of the indiscriminate use of pesticides on wildlife, particularly birds. This was a clear example of the unintended, deleterious effects of technology and started a global environmental movement, which unfortunately has been equated with the antitechnology movement by many people. In fact, the environmental movement is based on science, with ecology as a main source, which rejects the control over nature by technology and is sensitive to its unintended effects on nature. The issue of complex and interacting human and environmental problems was taken up by a group of scientists, educators, and business and political leaders, who called themselves the Club of Rome. They correctly perceived that unlimited economic growth would put humanity in trouble with the planet. In 1972, they produced a book, *Limits to Growth* [49], with a computer simulation of economic and population growth, which predicted that growth could not continue indefinitely because of the limitation of natural resources and other problems, which would lead to the collapse of the global system.

Concerns over the rights of the human family to a healthy and productive environment led the United Nations to convene the Conference on the Human Environment in 1972 and the subsequent formation of the UN World Commission on Environment and Development, chaired by Gro Harlem Brundtland, former prime minister of Norway. The commission produced a report entitled *Our Common Future*, also known as the *Brundtland Report* [50], in 1987. A subsequent UN Conference on Environment and Development (UNCED) in 1992, also known as the Rio de Janeiro Earth Summit, Rio Summit, Rio Conference, and Earth Summit [51], addressed the issues of unsustainable patterns of power production, water supply, and environmental degradation, among others. The summit gave rise to the declaration of Agenda 21, an action plan with regard to sustainable development. The UN Framework Convention on Climate Change (UNFCCC) is a treaty arising from the summit, with the objective to stabilize greenhouse gas levels in the atmosphere, with the Conference of the Parties (COP) as a major instrument to take actions and assess progress. Important landmarks include the adoption of the Kyoto Protocol in 1997, which set up internationally binding targets for greenhouse gas emissions, and the Paris Agreement in 2015, which set the limit for the global temperature increase to no more than 2°C. The summit also gave rise to important agreements, including the UN Convention on Biological Diversity dedicated to the conservation of biodiversity, sustainable use of its components, and fair sharing of benefits arising from use of genetic resources. In 2012, a Rio+20 Conference, or the UN Conference on Sustainable Development, was held with renewed commitment of the Earth Summit and broadened goals, namely achievement of a green economy, poverty eradication, and an institutional framework for sustainable development. In "The Future We Want," the outcome document of Rio+20, we are witness to a movement that started as an environmental one, broadening to embrace societal concerns, with the realization that efforts to achieve sustainable development must involve social and economic agenda as well as environmental ones.

## 10.3 Sustainable Development Goals

Human and environmental problems were the main concerns of the Millennium Summit in 2000, where world leaders adopted the UN Millennium Declaration, committing their nations to a global partnership to eradicate extreme hunger and poverty, achieve universal primary education, promote gender equality, reduce child mortality, improve maternal health, combat infectious diseases, ensure environmental sustainability, and achieve a global partnership for development. The Millennium Development Goals (MDGs), as they were called, were targeted to be realized in 2015. When that date approached, it was clear that the goals would not be achieved by many developing countries, although impressive progress was made in a number of them. The UN Sustainable Development Summit in 2015 made a renewed, and more comprehensive, plan to achieve development goals by the year 2030. In effect, the summit attempted to deal with concerns on sustainable development of social, economic, and environmental issues of the world. The Sustainable Development Goals (SDGs) cover 17 topics, with a total of 169 targets covering a broad range of development issues.

The 17 goals (Fig. 10.2) are as follows:

1. End poverty in all its forms everywhere.
2. End hunger, achieve food security and improved nutrition, and promote sustainable agriculture.
3. Ensure healthy lives and promote well-being for all at all ages.
4. Ensure inclusive and equitable quality education and promote lifelong learning opportunities for all.
5. Achieve gender equality and empower all women and girls.
6. Ensure availability and sustainable management of water and sanitation for all.
7. Ensure access to affordable, reliable, sustainable, and modern energy for all.

8. Promote sustained, inclusive and sustainable economic growth, full and productive employment, and decent work for all.
9. Build resilient infrastructure, promote inclusive and sustainable industrialization, and foster innovation.
10. Reduce inequality within and among countries.
11. Make cities and human settlements inclusive, safe, resilient, and sustainable.
12. Ensure sustainable consumption and production patterns.
13. Take urgent action to combat climate change and its impacts.
14. Conserve and sustainably use the oceans, seas, and marine resources for sustainable development.
15. Protect, restore, and promote sustainable use of terrestrial ecosystems, sustainably manage forests, combat desertification, halt and reverse land degradation, and halt biodiversity loss.
16. Promote peaceful and inclusive societies for sustainable development, provide access to justice for all, and build effective, accountable, and inclusive institutions at all levels.
17. Strengthen the means of implementation and revitalize the global partnership for sustainable development.

**Figure 10.2** Sustainable Development Goals adopted by the United Nations in 2015.

The SDGs are more numerous than the earlier MDGs and have a distinct set of targets under each goal. The approach of setting numerous goals and targets has been criticized by some as presenting the issues in silos, with little integration between them. The targets have also been criticized as both lacking integration and overlapping with one another, and using vague language rather than precise quantitative terms. In spite of these criticisms, the SDGs serve as a set of achievable goals toward sustainability agreed upon in general by the global community. They represent human aspirations and attempts to deal with both human and environmental problems. There is continuous assessment of achievements and problems at country levels, enabling cross-comparison between countries and encouraging countries to try to improve their performances measured against the goals and targets. There is still time before 2030 (Fig. 10.3), the final year for the goals, to improve both the definition of targets and the performances toward achieving the targets. Lack of progress toward the goals, or even backward slides, can still occur in some countries or societies, but they will not occur in isolation. There will be opportunities for the world community to work together to correct the situation and not let complete collapse occur as in isolated societies of the past.

**Figure 10.3** Awaited transformation by sustainable development.

Before we go on to scrutinize the individual SDGs, we should examine overall concepts that will enable us to reach such goals in total, together with surveying the status and future action needed. One overall concept that has been employed successfully in Thailand and other countries is called *sufficiency economy* [43]. As Box 10.1 shows, sufficiency thinking in this philosophy is not aiming for self-sufficiency in production and consumption but urges moderation, reasonableness, and building of immunity against adverse conditions in economic and related activities so as to achieve sustainability at both personal and community levels. Thailand's present status concerning sustainable development has been compiled in various aspects, ranging from the environment to the economy, the society, and culture, with suggested roles for various parties [52]. This was followed by a call to country-wide action in various areas [53], resonating the United Nations' message on "a call to action to change our world." These are examples of what individuals, groups, communities, and countries can do together to achieve a sustainable world.

## 10.4 The Role of Science toward Achieving Sustainable Development Goals

The 17 SDGs can be roughly divided into five groups. Goals 1–5 are about people, goals 6 and 12–15 are about our planet, goals 7–11 are about building sustainable prosperity, goal 16 is about peace, and goal 17 is about partnership. Altogether, they provide not only the final goals to be achieved but also goals that serve further as a means to achieve other goals. They should not be taken in isolation from one another, as criticized to be the silo approach, but should be woven together and be contributing toward each other. For example, getting rid of poverty will have uplifting effects on hunger, health, education, and progress toward sustainable cities, among others. The other way round, progress toward the other goals will have a positive feedback on the end of poverty.

Just as the goals should be taken as interrelated and not in isolation from one another, so should the moves toward achieving the goals be made as integrated approaches, with appropriate mixes between economic, social, environmental, community-based, financial, business, legal, scientific, and technological inputs, as appropriate. As a universal public good that empowers people to find the solutions they need, science and technology play a pivotal role in our attempts to achieve sustainable development. This was recognized by the UN secretary-general, who set up a Scientific Advisory Board to form recommendations on how to maximize the contributions of science and technology to the potential achievement of the SDGs. Here we will concentrate on the scientific and technological input toward such achievement, bearing in mind the importance of integrated approaches. We will also note that in some cases, scientific and technological advances may have adverse effects on attempts to reach the goals, and preventive or mitigative measures should be taken, wherever possible.

1. **End poverty in all its forms everywhere**: Ending poverty is an important part of the MDGs launched at the end of the last millennium, and by 2015, the poverty rate, or share of the population under a given poverty line, decreased by more than half, meeting the main target of the goal [54]. Poverty is not just lack of money but also lack of the means to fulfill reasonable human needs. Science and technology help in making effective livelihoods, resulting in earning and alleviation of monetary poverty. For example, small farmers can gain from information provided by satellite- and ground-based data on soil and water, plus agronomic and market information from extension services, so as to choose the best crops to grow and sell. Community-level industries can gain from upgrading of their products and better access to the market through access of information made possible by widely available Internet and mobile technologies. Planning of work and business and financial accounting of income and expenditure have also been much facilitated by such technologies. People can reduce their expenditure, such as by making natural pesticides rather than buying chemical ones, with the bonus of lesser environmental impact. Indigenous knowledge of the environment and local history helps in the promotion of sustainable tourism, which can be done at the village level. Renewable materials, such as fabrics and food items, can be produced for sale or for personal consumption.

Solar drying, cooling, and other uses of solar energy can help produce such items and bring comfort to homes without major expenditure.

Automation and other advances in technology can have a negative effect on employment and undermine efforts to end poverty. Good policies and measures are needed to prevent drastic displacement of labor and to ensure that the people have appropriate education and training for new knowledge and skills required in the changing employment landscape. Technology can be turned from a threat to a tool to upgrade and diversify earning, for example, farm mechanization that helps in productivity and frees up time for farmers to earn more from other employments.

2. **End hunger, achieve food security and improved nutrition, and promote sustainable agriculture**: The end of hunger is closely related to the end of poverty, and many approaches outlined for the latter are also relevant to the former. Here the approaches should be aimed at achieving sufficiency in feeding oneself and the family, with surpluses to be sold for financial income. Most families in developing countries have only small pieces of land and have to achieve food security, good nutrition, and sustainable agriculture with only limited means. Successful models have been shown for small-scale farms, with mixed and rotating crops, good water management, and good agricultural practice. These can provide enough to feed the families, with some excess products for income generation, utilizing appropriate varieties from breeding technologies. In many cases, supplementary nutrients are needed, such as iodine in areas far from the sea. In such cases, governments can come to the assistance of the community by providing needed nutrients at reasonable costs. Governments or local authorities can also assist farmers in many areas with problems with soil quality or water access through provision of appropriate technology and management.

   Commercial industries, enabled by technology, produce not only nutritious foods and drinks but also, by market demand, those that have low values or even detrimental effects on our health. Obesity is an increasing nutritional problem on the flipside of hunger, especially for people living in urban areas, and it needs to be targeted as well as hunger. This goal is therefore not just about the elimination of hunger but also about the provision of nutritious food and the best means to do so.

3. **Ensure healthy lives and promote well-being for all at all ages**: Freedom from illnesses and general wellness come from healthy living in healthy surroundings. Medical science and public health can promote wellness both through preventive measures, such as the provision of vaccines, and through the treatment of many diseases. Infectious diseases such as malaria, tuberculosis, dengue hemorrhagic fever, and AIDS are still affecting a major portion of the global population; however, major efforts are underway to eliminate these diseases. Drugs to fight diseases such as AIDS and hepatitis C, which used to be out of reach of poor people due to high prices, are now becoming more affordable through various programs with cooperation from pharmaceutical industries. However, a major threat is looming in the fight against infectious diseases since resistance to antibiotics and other drugs has increased, forcing the search for new remedies that are not vulnerable to resistance. New progress in precision medicine, defined as an emerging approach for disease treatment and prevention that takes into account individual variability in genes, environment, and lifestyle, will improve the lives of individuals. This approach for the public as a whole, with appropriate use of big data, will lead to far more effective public health measures in the future. For example, preventive measures can be made for groups of people with similar genetic susceptibility to diseases or sensitivity to certain environmental conditions. The key to achieving this goal is to promote a healthy lifestyle and a healthy environment for everyone, which depends on progress in other goals, such as freedom from hunger and sustainable cities for us to live in that have good sanitation and are free from pollution.

   Health technology is becoming increasingly expensive, and many people cannot afford it in their time of need. Pharmaceutical companies have the moral responsibility to make sure that their drugs are affordable for people in need. Governments need to make sure that the pricing of medical services by the commercial sector is appropriate and that there is adequate protection for people through appropriate health insurance policies or universal health care.

4. **Ensure inclusive and equitable quality education and promote lifelong learning opportunities for all**: Information technology is a powerful tool to provide quality education and promote lifelong learning opportunities for all. One major task is to develop suitable learning content, such as online courses and training materials. There has been major progress in the development of such online courses, but these are still mostly in English, and there is a need to have such courses in the languages of the developing countries. Another major problem is the availability of teachers who can perform both the role of traditional teachers in personal rapport with students and the link between students and the online materials. Furthermore, quality education should cultivate essential 21st-century skills in the students so that they can become good citizens, contributing to both local and global communities in an increasingly connected world. Education with emphasis on science, technology, engineering, and mathematics will equip future students with knowledge and technical skills required for the society of the future. This is especially important considering the fact that jobs in the future will change with the introduction of new technologies and businesses and lifelong employment in performing the same task will become rare.

   People do not get education just from the classroom but also from all sources in the environment, including immediate families, communities, and various media. Indeed, research has shown that investment in people has the best return for preschool children, with education input mainly from the families and communities. As for other age groups, technology embedded in cell phones and the Internet, together with mass media and social media, has helped in mass education tremendously, but it has also made it easy to spread untruths, immoral messages, and miseducation in general. To achieve success, this SDG should take account of not only formal education but also this nonformal information.

5. **Achieve gender equality and empower all women and girls**: Women and girls in most societies suffer from inequality and discrimination both in life and at work. Empowering them should be an important goal in sustainable development, as they can fulfill their potential roles as half of humanity. In the past few decades, science has made it possible for women and their families to reduce the burden of large families and unwanted conceptions. Women have been liberated from the traditional role of merely belonging to the homes. Science has shown that women are comparable to men in various capabilities, except for those that depend on different physical characteristics. Given opportunities, women can rise to the top of career ladders and be effective entrepreneurs, sportspersons, political leaders, and scientists. Science has helped women assume dual roles, both doing their jobs in the outside world and being wives and mothers at the same time. The Internet and various gadgets derived from science have enabled women to work in the homes and cut the time and labor needed for household jobs. However, only a small portion of women worldwide enjoy the new opportunities given to them by technology, since most are still poor, with little education. Importantly, most societies still hold on to the traditional roles for women and girls that do not allow them to achieve gender equality. For this goal to be achieved, not only must other goals also progress hand in hand with it, but societal attitudes also have to change to allow more opportunities for women.
6. **Ensure availability and sustainable management of water and sanitation for all**: Management of water includes management of water availability and access, threats of floods, and quality of water. The first two areas are domains of the science called hydrology. Remote sensing and ground survey technologies allow planning of water management at the local level as well as on a large scale. Droughts and floods can be controlled by regulation of water reserves, inflow and

outflow. Many countries build large reservoirs and dams to regulate water flows of rivers. Water flow can be used to generate power from these hydroelectric dams. However, this is often associated with problems of people upstream of the dams, who need to be relocated due to high water levels, and downstream people, who are prone to periodic changes in water levels due to flow regulation. Furthermore, large dams in tropical countries, such as the Aswan Dam in Egypt, become breeding grounds for snails and other sources of parasitic diseases such as schistosomiasis (caused by blood flukes). Appropriate technology is available for making small dams and weirs, either singly or in a series. Irrigation from reservoirs and rivers provides water for agriculture and household purposes, as well as helping regulate amounts of water to appropriate levels. As for issues concerning the quality of water, sanitary science is instrumental in providing clean water for consumption and other purposes. Water-borne diseases, including diarrhea, leptospirosis, and skin infections, are controlled with the help of this science. Again, the success of such control depends on the progress of other goals, including end of poverty, freedom from hunger, and creation of sustainable cities.

7. **Ensure access to affordable, reliable, sustainable, and modern energy for all**: Renewable energies like solar energy, wind, and biofuels, are sometimes called nonconventional or alternative energies because of the fact that they traditionally form only a minor part of our energy supply, the bulk of which comes from fossil fuels. They are environment friendly, unlike fossil fuels, which produce much more greenhouse gases and other environmental pollutants. The situation is changing dramatically, thanks to the science of renewable energy. Solar panels are composed of photovoltaic cells using silicon semiconductors to convert light energy into electricity. These cells have been improved continuously, and new materials and methods of fabrication have resulted in increasingly higher efficiency and lower cost per unit power. Solar panels on rooftops can produce electricity for household use and promise to have excess energy produced that can be fed into

central grids, reversing the conventional direction of energy supply. Other means of solar energy conversion besides photovoltaic cells are also being developed. At the community level, solar energy provides an appropriate source of energy for drying and cooling. Wind energy is being harnessed at a large scale with increased efficiency, and although it has environmental effects like noise and use of space, the effects are far less than those of fossil fuels with carbon dioxide as the main pollutant. Biofuels like biodiesel and bioethanol are produced mainly from agricultural and waste products, and although they also produce carbon dioxide on consumption, their origins involve conversion of carbon dioxide into biological products. Fast-growing algae and microbes are potentially attractive sources of biofuels, the development of which relies on genetic and other sciences.

The technology of hydraulic fracturing, or fracking, is making it possible to extract oil and gases from deep underground. Other technologies are making it more economic to pump oil and gas from the deep seabed. These help increase the supply of fossil fuels, which are the main sources of world energy. However, these developments raise questions about environmental effects and may hamper the search for alternative energy sources that are kinder to the environment by producing less greenhouse gases and other pollutants.

8. **Promote sustained, inclusive, and sustainable economic growth, full and productive employment, and decent work for all**: Technology and innovation are increasingly important in achieving economic growth. This is especially true for the private sector, competition in which has been repeatedly shown to depend on superior technology. New technologies tend to be disruptive, that is, they disrupt the markets with the introduction of new products and services, which are judged by consumers to be better, more powerful, and more convenient, and displace the old products and services. The displacement of photography that depends on chemical processes by digital photography is an example of the effect of disruptive technology. The rise of the electric car to replace cars run by the internal combustion engine is

another example. In the service sector, the use of block-chain technology is replacing age-old bank-mediated processes in commercial transactions. The disruptive technologies will, in time, permeate from commercial sectors to the public in general, as seen in mobile telephony and e-commerce. However, sustained, inclusive, and sustainable economic growth need not come from disruptive technologies alone. Indeed, conventional technologies, used in ways that promote sustainability, can contribute significantly to achieving this goal. Technologies for sustainable production of food and medicine originally derived from the forests, such as tissue culture, domestication, and plantation for harvest of these products, can result in economic growth without destroying biodiversity. Technologies for making long-lasting insecticide-treated bed nets are derived from regular scientific knowledge, and yet, with appropriate policy for their utilization in various countries, they have made a big impact in the effort to turn back malaria and other insect-borne diseases.

Many technologies are disruptive, not only to markets, but also to people whose jobs and livelihoods depend on the disrupted production and services. Agricultural machineries are more efficient than human labor, e-mails have made the conventional service of the post office obsolete, and the jobs of bank tellers are done faster and more economically by ATMs and online banking services. These are only a few examples of disruption by technology, and there are many more on the horizon. To be sustainable, promotion of economic growth should take the human labor factor into account. People should have training so as to be capable of doing new tasks and not be stuck with outdated skills. This highlights the importance of the goal of promoting quality education and lifelong learning opportunities.

9. **Build resilient infrastructure, promote inclusive and sustainable industrialization, and foster innovation**: Infrastructure is the basic framework a society needs in order to function properly, including physical structures such as road and rail systems, airports, harbors, water supply, and waste management. It also includes equipment

and facilities for information communication, education, science, technology, and innovation. Resilient infrastructure, important for sustainable societies, is composed of structures and systems that can withstand the wear and tear of usage and the hazards of natural or man-made disasters such as storms, floods, accidents, and terrorism. Such infrastructure is needed for a well-functioning society, in which inclusive and sustainable industrialization can be built. Science largely provides the means and materials for building and maintaining such infrastructure, ranging from construction materials to the global Internet system. It can help promote inclusive and sustainable industrialization based on such resilient infrastructure, which is a more desirable outcome than the short-lived industrialization of the past. The resilient infrastructure needs to support flexible manufacturing and services, which have to evolve together with changing demands. In this respect, infrastructure for innovation, science, and technology is an important part of the resilient infrastructure and needs to be nurtured constantly by the government and the public so that both the public and the private sector will be able to develop inclusive and sustainable industrialization and services. Public expenditure must take account of building the innovation infrastructure in the form of communication systems, research facilities, international personnel, and investment in scientific infrastructure. The private sector should be encouraged to perform innovation through availability of such infrastructure and appropriate tax and incentive measures.

10. **Reduce inequality within and among countries**: Inequality within and among countries can be reduced by a number of international, national, and community actions. These actions include policy and support programs to create jobs and businesses at the grassroots level, further public education, and provide adequate health and social welfare, backed by appropriate tax and other policies, such as those encouraging microfinance for community-level businesses. Science plays a big role in enabling people at the base of the pyramid to help themselves find suitable livelihoods and improve their

standard of living. For example, people can obtain education and information for taking care of their health, starting a home business, and keeping account of their income and expenses through computers, mobile devices, and the Internet. Local food industries, consumer product industries, arts and craft, and ecotourism can be launched at the community level with the help of these media plus local experts with good knowledge of "folk science," much of which is based on universal science with local examples and experience added.

At the international level, development assistance from advanced countries can play a big role in reducing inequalities among and within countries. A major part of such assistance involves helping people develop good agricultural practices, good services, and good local industries and businesses and improve their daily lives, much of which have science as the guiding principle. Water management, for example, relies on knowledge of seasonal sources of water, its storage, and its mobilization, utilizing satellites and ground-based hydrology information. Access to clean water needs a basic knowledge of filtration and sanitation. Finding and conserving energy for household use requires knowledge of rural energy, including solar and biobased energy. Construction of houses can be done economically and with higher quality with scientific knowledge of energy conservation, appropriate housing materials, and sanitary considerations. Universal scientific principles can be blended with local knowledge and expertise to give the best outcome.

11. **Make cities and human settlements inclusive, safe, resilient, and sustainable**: Science can help make sustainable cities and human settlements through its contribution to good planning of towns, forests, and parks, using the principles of architecture, built environment, and, what is collectively called, regional science. Inclusive, safe, resilient, and sustainable cities are those with a good transport system, utilities, good energy supply, clean air, good waste recycling, and other physical infrastructures. For example, traffic management can be considerably improved through a computerized system, taking into account the amount of

traffic, routes, and flow regulation, together with information provided to road users. In sustainable cities and settlements, areas for housing, shopping, offices, and manufacturing facilities are distinct but interconnected. People can travel between homes and workplaces without having to spend too much time traveling or to pay excessively for transport. There are parks, walkways, cycling routes, and places for relaxation, entertainment, and cultural activities accessible to the public at large. Furthermore, education and health facilities are widely available and affordable to people in the cities and settlements without having to travel long distances.

As urbanization—a global trend—progresses, cities will face growing pressure on space, infrastructure, and services. Of special concern is the pressure on congested areas of dwelling as seen, for example, in Rio de Janeiro, Kolkata, or Bangkok. Good public policies and measures are needed, either to improve present conditions or to prevent them from getting worse and, where possible, to relocate the inhabitants. Similar problems are faced by people in areas reserved as national parks and other public spaces. Science can play a part in the planning of relocation through the use of geographic information systems and knowledge concerning planned terrains. However, this is a sensitive issue requiring consideration from humanitarian and many other aspects, not just science. An integrated approach involving many areas of work is required for solutions to the problems of cities and human settlements in general.

12. **Ensure sustainable consumption and production patterns**: The world does not have unlimited resources. As the population grows, there is increasing danger of reaching the planetary boundaries, both in providing resources for consumption and in reaching the limits of a sustainable world in environmental change and keeping social harmony. Excessive consumption and production that encourages or attempts to satisfy the increased consumption contribute further to sustainability problems already caused by population increase. As discussed in Chapter 7, sustainable consumption and production patterns are prerequisites for both present

and future societies. The 3R principle is appropriate for our approach to sustainable consumption and production: reduce, reuse, and recycle. Science can help us achieve the goals of the 3R principle. We can reduce consumption and production by optimizing the use of products and materials, reducing waste or excess, and synchronizing the production with consumption so as to reduce excess supply. We can reuse various materials and products such as clothes, papers, and parts of cars and machines, sometimes after transforming them to other products. As for recycling, this is now a flourishing industry, ranging from recycling of plastic and glass bottles to recycling of fabric and household waste.

Sustainable consumption and production patterns have been advocated by many influential thinkers, including the late King Bhumibol Adulyadej of Thailand. His Sufficiency Economy Philosophy (see Box 10.1) is a set of social, economic, and environmental guidelines for people to live in harmony with nature and society through moderation, reasonableness, and building of immunity against adverse changes [43]. The UN Partnerships for SDGs have included the principles of the Sufficiency Economy Philosophy as an innovative approach toward this and other goals of sustainable development.

**Box 10.1** Sufficiency Economy Philosophy as a Contribution toward a Sustainable World

We are living in a potentially unsustainable world, reaching or overstepping planetary boundaries beyond which it will be difficult or impossible for natural adjustment to occur for the environment and human society to avoid a catastrophe. Mahatma Gandhi foresaw our current problems when he said that the earth provides enough to satisfy our needs but not our greed. It appears that greed is a main force driving us toward unsustainability. Global warming from our excessive use of energy has pushed the atmospheric temperature up with foreseen consequential disastrous effects on crop yields, water supplies, and sea levels. We are harvesting marine resources to the extent that exceeds the capacity of natural replenishment. We are polluting the land, the seas, and the air with chemicals that threaten biodiversity and human health. A substantial part of the problems is due to excessive consumption and production patterns, noted in the elaboration of goal 12 of the SDGs. Such excesses in consumption and production, spurred on partly by uncontrolled market forces and our own unchecked greed, are unnecessary and proving to be dangerous in the long term. Fortunately, we can adapt our ways of life to avert this danger. Throughout history, when times were hard, and even today in less wealthy countries, people could live normal lives happily with fewer resources. Studies have shown that once basic needs are satisfied, further unlimited consumption does not make people happier. Models for sustainable consumption and production patterns can, therefore, be expected to be found in relatively less wealthy countries and communities, which manage to find optimal use of their limited resources.

Take Thailand as an example. The Sufficiency Economy Philosophy, elaborated by His Majesty the late King Bhumibol Adulyadej, urges people to take sensible and wise decisions in running their lives on the basis of three major principles: moderation, reasonableness, and building immunity and resilience to adverse changes. The concept of moderation is in line with antigreed teachings of major religions, balancing between need and extravagance. Reasonableness is the adoption of fairness and logic in thinking and action. Immunity is the ability to deal with changes and risks so as to achieve resilience to adverse events through knowledge and exercise of prudence. Rather than aiming for self-sufficiency in the sense of isolation from world production and consumption, the philosophy encourages people to rely on themselves in making decisions based on these three principles. The philosophy can be adopted by individuals, families, communities, and businesses, as well as at the national level. This philosophy was initially directed to the population of Thailand, a middle-income country that has to deal with forces of globalization, but is clearly in line with our attempts to achieve sustainable global consumption and production patterns. It is also in line with models of sustainable leadership of enterprises in Europe, identified as "honeybee" leadership in contrast with "locust" leadership [43]. The former takes a long-term, stakeholder approach, taking all stakeholders into account, in contrast with the latter, which emphasizes short-term wins for investors at the expense of other stakeholders.

Apart from contributing to the goal of sustainable consumption and production patterns, the Sufficiency Economy Principle is also applicable to other goals, including sustainable management of the world's ocean and land-based resources and promotion of sustainable and inclusive economic growth. Thailand is sharing the Sufficiency Economy Principle and putting it into practice with other countries in what is known as the South–South Cooperation.

13. **Take urgent action to combat climate change and its impacts**: Of all the factors threatening sustainability of the earth and human society, climate change stands out as the most important. Over the past century, there has been a gradual increase in the average temperature of the earth's atmosphere by around 1°C. Global warming is due to various causes, but the predominant cause for the recent rise in global temperature is human activities, predominantly release of carbon dioxide from burning fuel for industrial and other activities. Svante Arrhenius, a Swedish chemist, predicted correctly more than a century ago that doubling of carbon dioxide in the atmosphere would cause a mean temperature rise of around 5°C. The level of this gas in the atmosphere has risen from around 300 parts per million (ppm) in the past century to 400 ppm today. Apart from global warming, oceans also undergo acidification due to absorption of the gas, leading to deleterious effects on coral reefs and marine life in general. Other gases from human activities causing global warming, collectively called greenhouse gases, include methane and nitrous oxide. The total emissions of greenhouse gases are increasing because of increased energy demand for industrial and other activities in advanced countries and have recently intensified by the rise of countries such as China. The Rio Earth Summit in 1992 already realized the problems of climate change and agreed to stabilize greenhouse gas concentrations at a level that would allow ecosystems to adapt naturally to climate change. However, this stabilization has not occurred, and it has been estimated that if we go on with the business-as-usual activities, the global temperature rise could be as much as 4°C–7°C by the end of the century. Atmospheric temperature rise causes wide-ranging effects, including decreased crop yield, reduced biodiversity, reduced freshwater availability, rise in ocean levels, and extreme weather events.

The UN Climate Conference in Paris in 2015, with 195 nations participating, agreed to combat climate change by taking action to limit temperature rise in the future to no more than 2°C over the preindustrial levels and even to no more than 1.5°C, if possible. This would require mitigation and adaptation measures by all parties, from governments to businesses to individuals. New industrial and household processes would need to be modified to become more efficient and use less carbon-based energy. Cars and other means of transport, for example, should become more efficient and rely more on solar and other alternative sources of fuel than the present fossil fuels. Buildings need to be upgraded so as to use less energy-intensive materials in their construction and to require less energy in heating, cooling, and other services. The efficiency of power plants needs to be improved, and alternative sources of energy that release minimal or negligible amounts of greenhouse gases need to be developed. Actions are needed to stop deforestation and increase reforestation. The means to capture carbon at the sources of release or from the atmosphere also need to be developed. These are some of the contributions that science and technology can make to help achieve the target of 1.5°C–2°C set by the Paris Agreement.

14. **Conserve and sustainably use the oceans, seas, and marine resources for sustainable development**: Covering 71% of the earth's surface, the oceans and the seas provide numerous resources without which we would not be able to survive. The oceans, seas, and marine resources have been overused and abused by humans, especially with increased demands for food, energy, minerals, trade, transport, and tourism. Unlike land, where food resources are now mostly obtained by agriculture, food resources from the oceans and seas are still mostly secured through hunting from nature. Technology has been responsible for increased catches through the use of modern fishing vessels that enable hunters to go far and deep. However, using technology to increase the efficiency of exploiting marine resources is not sustainable in the long run. The situation is worsened by illegal, unreported, and unregulated (IUU) fishing. It has been estimated that the

harvest of fish in the tropics will be reduced by 40% by 2050. Clearly, we need to conserve and sustainably use marine resources. Science and technology can be used to achieve this purpose by concentrating more on conservation and sustainable production rather than hunting. Aquaculture and mariculture, or farming of fish and other aquatic resources in freshwater or seawater, are promising areas of science that need to be developed sustainably. Care should be taken to avoid conflicts between aquaculture and environmental conservation. A case in point is shrimp farming along shore lines, which threatens mangrove forest areas. As for deep-sea fishing, technologies for regulation and monitoring of vessels in the seas and oceans, including the size of catches, can help toward sustainability of the resources. Technologies for surveying and monitoring marine species like tuna should also be deployed for their sustainability as food sources. Other vulnerable marine ecosystems, including coral reefs and sea grasses, should also be monitored and conserved with the help of marine science and technology.

International and national regulations are also needed in the search for, and exploitation of, energy and mineral resources in the oceans and seas. In the past, technology has been developed to explore, identify, and extract such resources, with little or no concern for environmental effects or sustainability. New science and technology need to aim at sustainability of such practices, not just commercial values.

As a consequence of global warming, new sea and ocean routes will become available for marine traffic, for example, through polar routes. International plans and agreements should be prepared, with the help of technology, to make good and sustainable use of such new routes in the future.

15. **Protect, restore, and promote sustainable use of terrestrial ecosystems, sustainably manage forests, combat desertification, halt and reverse land degradation, and halt biodiversity loss**: Terrestrial ecosystems include all life forms and their surroundings on land, including forests, deserts, permafrost, and human habitats. Ecosystems are subject to natural change but are increasingly threatened by

humans through encroachment on forests, land degradation resulting in desertification, and global warming with the destruction of the polar ice caps. Population growth puts further pressure on the ecosystems. Deforestation occurs as a result of both expansion of human settlement areas and the exploitation of forest products for commercial purposes with consequences such as reduction of rainfall, inability of the land to retain rainwater, floods, soil erosion, and loss of biodiversity. Human activity has resulted in extinction of many species, which has become so severe that many have signaled recent events as the beginning of what is called the Sixth Great Extinction, or Anthropocene Extinction. Loss of forests also results in decreased absorption of carbon dioxide by plants, exacerbating global warming. We therefore need to protect and restore our terrestrial ecosystems as a matter of urgency. Land use and land conservation policies need to be developed, with the means to enforce laws and regulations. Zoning of land areas as forests, agro-ecological areas, and areas for sustainable human use should be considered. Earth observation through remote sensing, ground surveys, and other geographic techniques can help in enforcing such policies and developing plans that take into account both human needs and needs for conservation. Land fertility can be sustained by a suitable agronomic practice such as the use of optimal crops for the land and crop rotation, biofertilizers, and suitable pest management techniques. Soil erosion can be prevented by suitable plants, for example, Vetiver grass with long roots holding the soil together. Degraded land can be reclaimed and restored through a variety of means, such as improving irrigation, changing soil acidity, and introducing appropriate vegetation. Reforestation of degraded land can be achieved in many cases with the incentive of earning carbon credits that can be sold to customers, such as companies that want to reduce their carbon footprints from their greenhouse gas emissions.

16. **Promote peaceful and inclusive societies for sustainable development, provide access to justice for all, and build effective, accountable, and inclusive institutions at all**

**levels**: This goal is about building peace, inclusiveness, and justice in societies for sustainable development. Social science and other areas of science, including psychology, are important for the understanding of conflict resolution and peace. Other areas of science provide powerful tools against acts of violence and fraud, for example, deoxyribonucleic acid (DNA) analysis to identify criminals and block-chain technology to provide transparency and traceability of financial transactions and other products. Information and surveillance technologies provide tools for protection of victims of violence, terrorism, and human trafficking and against arms trafficking. The Internet and various media enabled by technology provide free access to information pertaining to human rights, laws, and justice for all people irrespective of their background or financial status. Through such technologies, people can also come together to form civil societies, pressure groups, or other organizations to protect their civil liberty and rights. In general, science provides enabling tools for building effective, accountable, and inclusive institutions at various levels, and it remains for people and governments to use these tools together with other means to achieve their targets.

17. **Strengthen the means of implementation and revitalize the global partnership for sustainable development**: This goal is about building a global partnership that is crucial to the success of our efforts to achieve all the goals for sustainable development. A global partnership for sustainable development has been built up over the past few decades and is now in the crucial stage in the attempt to reach defined goals. Science plays a general enabling role for the various efforts to meet this goal. It provides good examples of partnerships in research and development. Large-scale projects in science, like the Human Genome Project of the past century, programs to explore space and deep oceans, and programs to investigate fundamental particles, all depend on goodwill and active partnerships between various countries. We can learn from the success of these programs and build diplomacy based on science, which can be used as examples of global partnerships. The Internet and other tools for

partnership development were developed first as scientific projects, which then expanded to other areas of human communication. In this respect, it is important that such tools be available to people in various parts of the world, including those in poor and developing countries, so as to enable them to be integral parts of global partnership.

## 10.5 Sustainability Science as a Key to Sustainable Development

We have seen that various areas of science can contribute toward achievement of different goals of sustainable development. These are the results of application of these specific areas of science to distinct goals, providing the bottom-up approach essential for solving discrete problems of development, including poverty, gender equality, sustainable cities, and climate change. However, in addition to this bottom-up approach, we need a broad overall approach to the problems of sustainable development. The various goals of sustainable development are linked together: success in achieving a particular goal depends crucially on progress toward other goals. To achieve lasting peace, the world needs to be free of poverty and hunger. To make progress in slowing climate change, we need sustainable cities and industries and the development of renewable energy, as well as conservation of resources on land and in the seas. These, in turn, depend on sustainable consumption and production. There is therefore a need to understand the chains of global events, both natural and human-factored, and their consequences and develop potential approaches to the problems on a broad intersectoral level. These approaches can be aimed at prevention, mitigation, or adaptation in the face of the chains of problems and their effects on one another.

Over the past two decades, a new science has emerged as a systemic study of the environment, human societies, and their interactions, with the aim of attaining sustainability both of the planet and of its human inhabitants. Sustainability science, as it is called, examines the dynamics of human–environment systems and the stability of their interactions, as well as interventions that promote sustainability in particular places and contexts. It is a holistic study combining various areas of natural science (e.g., climate science, ecology, oceanography, mathematical modeling, biology, and health sciences) with social science and humanities (e.g., development studies, economics, culture, and history). The unifying theme for these various areas of study is to understand sustainability and factors that promote or decrease it, both in the overall global and the local context. It should be an important contribution in the effort to achieve the stated SDGs and beyond.

# Chapter 11

# Moving Beyond Sustainability

*We have always held to the hope, the belief, the conviction, that there is a better life, a better world, beyond the horizon.*

—Franklin D. Roosevelt, *Address on Hemisphere Defense*, Dayton, Ohio, 1940

In the long run, we need to move further beyond efforts to achieve sustainability. No matter how successful we are in the attempt to achieve sustainability, we need to face the future as best as we can. Furthermore, desirable though it may be, sustainability implies maintenance of the existing state of affairs concerning the society and the environment, while we look further to their improvement and guarantee against future deterioration. For ecosystems, we should look beyond conservation to their productive evolution and enhancement of their diversity. For human societies, we should look beyond the absence of poverty, wars, and other scourges to the days when people can live in happiness without those fears. Science is an important element in the efforts to move beyond sustainability. In the long run, science should also help us in meeting our destiny as human beings on this planet and elsewhere.

*Sparks from the Spirit: From Science to Innovation, Development, and Sustainability*
Yongyuth Yuthavong

ISBN 978-981-4774-57-4 (Hardcover), 978-1-315-14599-0 (eBook)
www.panstanford.com

## 11.1 Are We Moving Toward or Away from Sustainability?

In a scenario based on modeling, researchers from the Universities of Maryland and Minnesota predicted that our current civilization is heading to a collapse due to its inherent instability unless rational policies regarding social inequality and environmental exploitation are adopted in time [55]. Technology would not be able to intervene to avoid the collapse and, indeed, would make the collapse more likely because it would increase the efficiency of extraction of resources from the environment. Although the study has been criticized for various drawbacks in the model and assumptions, it raised a legitimate point that unless we correct the concerns raised in time, the efforts toward sustainable development will be doomed to failure.

Yet, looking back over the recent past, there are some good signs for the future, especially now that the community of nations has come together to try to achieve the Sustainable Development Goals (SDGs). Although the preceding efforts of the Millennium Development Goals (MDGs) that measured progress between 1990 and 2015 did not achieve complete success, some targets were met or surpassed [54], such as halving the proportion of people living below the poverty line and halving the proportion of people without access to improved drinking water. Others were narrowly missed but showed impressive results, including reducing the proportion of people suffering from hunger (down from 23% to 13% between 1990 and 2015), reducing the proportion of out-of-school children of primary school age (down from 100 million to 57 million between 2000 and 2015), reducing the mortality of children and mothers, and halting the spread of HIV/AIDS. Monitoring of the progress of the SDGs has been made through the setting up of the SDG Index and dashboards for individual countries, and we should know their progress in attempting to meet the various goals and come together to solve the problems. We should follow such progress, or lack thereof, in meeting the goals of sustainable development from these and other indicators.

## 11.2 Beyond 2030

Beyond 2030, and even before that time, we should know whether we have met or are on the way to meeting the goals of sustainable development. We should not only find out how we are doing both globally and in our own immediate community and environment but also try to find ways to improve our performance. The SDG dashboards [56] for various countries give reports on how they are doing for each of the goals, with tags in green, yellow, or red. We should try to move from red toward green and identify where our weak points are and try to overcome them. These efforts should be continuous and started as early as possible. Individuals as well as organizations, communities, and governments can all make a difference in the collective global effort to attain sustainability.

Improvements can always be made to our performance, no matter whether we are scoring red, yellow, or green. In most cases, such improvements need many parties and many approaches, including social, economic, and environmental ones. Science is likely to be a key element in these efforts, as we already saw its potential contribution to various goals. Some interventions may need research, such as the capture of carbon dioxide at the source of release or from the atmosphere in the attempt to reduce greenhouse gases or enhancement of agricultural productivity through environment-friendly means. Others need applications of already known scientific principles or technological development, such as access to the Internet at the village level or promotion of nutritious foods and healthy ways of life.

## 11.3 Beyond Sustainability

Important though sustainability is, as has been emphasized throughout this book, it is but one aspect of a happy life and a happy planet. Indeed, "sustainability," in the sense of the ability to last, is a neutral term that denotes durability without the context of desirability. Some people have evoked other terms as desirable beyond sustainability. Hence, beyond the goals of ending extreme poverty and hunger, we should look forward to gaining a livable income and nutritious foods for all or most of the world population. We need to go further from freedom from diseases to achievement of wellness through integrated measures from womb to tomb. "Livability" is a term that denotes the quality of life worth living, with a healthy environment, economic security, social cohesion and equity, educational opportunities, and cultural and recreational experiences. "Thrivability" is the ability to grow and flourish in a healthy and vigorous way. "Resilience" is the ability to spring back or recover to a healthy state again after difficulties. These are some of the desirable qualities we should look for, which go beyond simple sustainability.

We also tend to view sustainability as something that requires perpetual human efforts to realize. For the environment, we tend to view nature as something that is continually violated by humans, and our efforts toward environmental sustainability are directed to conservation and prevention of calamitous effects like mass extinction and ocean acidification. These are, indeed, the main issues that we need to solve in the efforts to gain sustainability. However, in the long run, we need to return to the concept that we are merely a part of nature, as already made clear to us in the teachings of Taoism and many philosophers of the past. By behaving as a part of nature rather than as an intruder, we can look forward to self-sustaining and self-reinforcing development rather than development that needs to be continuously sustained by extra efforts.

## 11.4 Sustainability and Happiness

Sustainability should be a main factor leading to happiness of the society and of people (Fig. 11.1). Happiness is a state of mind denoting contentment and joy. It is a human feeling, a consequence of having fulfilled long-cherished hopes and dreams for some people or simply just playing games or enjoying the food they like for some others. Since ancient times, philosophers and religious leaders have put happiness as the central concern of human aspirations. A school of thought known as utilitarianism, led by 18th- and 19th-century British philosophers Jeremy Bentham and John Stuart Mill, states that all actions should be directed toward achieving the greatest amount of happiness for the greatest number of people. In modern times, many people have argued that happiness should be the real goal of economics. His Majesty King Jigme Singye Wangchuk of Bhutan pioneered the concept of gross national happiness, which measures the happiness of a country not only on the basis of wealth but also on the basis of other factors, including culture, religion, good governance, and sustainable development. Although there are some difficulties in the subjective nature of a few indices of happiness, this concept has gained recognition in many countries and has been the foundation for other international measures of happiness.

**Figure 11.1** Beyond sustainability lies happiness.

The *World Happiness Report* [57], published by the UN Sustainable Development Solutions Network, gives an analysis of happiness and a list of countries ranked according to data compiled from polls on six different topics. The topics are gross domestic product (GDP) per capita, social support, healthy life expectancy, freedom to make life choices, generosity, and perceptions of corruption. People are asked how they evaluate life in their countries on a scale of 0 to 10 compared to a hypothetical worst country, Dystopia, with a score of 0. Not surprisingly, countries that rank high are mostly developed countries, but some developing countries also score well. Many of the topics contributing to the happiness score are similar to the goals of sustainable development. Another measure of happiness takes the cost to the environment into account. The Happy Planet Index (HPI) [58] combines four elements to arrive at the scores: a sense of well-being, life expectancy, inequality of people within the country, and the ecological footprint. The HPI reflects the average years of happy life produced by a given society, nation, or group of nations per unit of planetary resources consumed, roughly representing the efficiency with which countries convert the earth's finite resources into well-being experienced by their citizens. Countries with top scores include many developing countries, mainly due to their relatively lighter ecological footprint.

These and other attempts to measure happiness have one common feature: happiness is linked to sustainability. Although individuals may differ with respect to attainment of happiness, in general some basic features of a good life must be present. To achieve happiness, we must at least have a decent social life based on our families and communities. We should be free from poverty, hunger, diseases, wars, and other social disturbances. We should have access to basic amenities like water and energy. We should have a good environment, both where we live and where we go for leisure. We should have basic human rights and freedom in the society we live in. Our society should be fair and just to everyone and be free from corruption and other crimes. All these basic requirements are in the SDGs, which we hope to achieve, not only for ourselves, but also for future generations. Admittedly, some people may still not feel happy even after they have achieved these basic conditions, while others may feel happy even though they lack them. In general, however, we can say that the SDGs are compatible with, and prerequisites for, a generally happy world population. As a major tool for achieving SDGs, science is therefore contributing to achieving happiness also. Apart from this general contribution, science is also helping us in the quest for happiness through the relatively new, specific area of "happiness science." This field of science, still to be further developed and integrated into mainstream science and other disciplines, involves the study of psychology in the positive sense. Positive psychology differs from mainstream psychology, which is concerned with understanding the human mind and repair measures when the mind is affected by illness. In contrast, positive psychology deals with understanding and building optimism, joy, resilience, and happiness.

## 11.5 Sparks for the Future

As we look to the future, we anticipate that the sparks from the spirit of science will continue to play a major role, perhaps even more so than in the past, toward innovation, development, and sustainability. We can even anticipate the role of science in the "postsustainability" era, when hopefully the problems of our own survival and that of the planet have been successfully solved. We can take a look to the future of human beings as a species on the planet and—in the long run—even beyond this planet.

### 11.5.1 Sharing Life with Artificial Intelligence

Our future will see a vastly increased role of artificial intelligence and associated devices ranging from robots to self-driving cars, toys, and household gadgets. The "Internet of Things" is already expanding connections of people to people, people to machines, and machines to machines, all of which increasingly depend on artificial intelligence systems connected through the Internet and other means of information and communication technologies. Our future life will see these connections expanded as machines become more intelligent. The cell phone, which has already greatly changed the way we communicate and conduct our daily lives in general, will likely become a remote control extension of ourselves, through which we run our lives or which runs our lives for us, depending on the point of view. Intelligent robots will do various jobs for us, ranging from industrial production to medical surgery to household work. Artificial intelligence will not only interact with us in the real world but also help create augmented reality, a hybrid between the real and computer-generated worlds, and virtual reality, where we experience virtual worlds that earlier could exist only in our fantasies.

The capacity of artificial intelligence is expanding at an increasing pace, making many experts predict that we are on the verge of *singularity*, a point where artificial intelligence surpasses all human intelligence (see, for example, Refs. [59, 60]). Predictions for the arrival of that point range around the middle of this century or even earlier. If, and when, that point arrives, we have to be prepared to live with the superior artificial intelligence that we created. Many pessimistic scenarios have been created, both in science fictions and in warnings by scholars, on how artificial intelligence, in the form of robots and other systems that we created, can be dangerous to our security and even our survival. For example, we can imagine a scenario where artificial intelligence is programmed autonomously or by humans to produce superweapons that cannot be neutralized. Another scenario is when artificial intelligence, with its own system of logic, takes control of human affairs from education to job assignment and lifestyles.

Artificial intelligence has undoubtedly been an important spark from the spirit of science. But how do we prevent the spark from igniting a fire that we cannot control? The key to our ability to share and improve our life with artificial intelligence without the dangers of being its victim is perhaps the fact that we as human beings not only have intelligence but also senses of ethics, justice, governance, beauty, and other characters of humanity. These hallmarks of humanity are related to, but distinct from, intelligence as can be acquired by computers. They are qualitative in nature and not well simulated by artificial intelligence, which relies mostly on quantitative and computable data. Although machines can be programmed to learn about ethical and other "human" values, up to now they are still primitive compared to us. The ethics of people-to-machine and machine-to-machine interactions should be well thought out by and under the control of humans, not left to autonomous decisions of artificial intelligence. As briefly discussed in Chapter 9, codes of ethics need to be developed for human–robot interactions concerning safety, human dignity, and privacy, as well as to cover the soft side of human nature, such as emotions and the liability to make errors. These codes of ethics should be built into programs for artificial intelligence as well as in the awareness of humans responsible for development of artificial intelligence, with the principle that the final judgment is under human control. Indeed, some have argued that artificial intelligence should not be developed for certain jobs such as those concerning ethical or moral judgment concerning safety, security, or life-and-death issues. The decision on medical intervention for serious illnesses, for example, should not be that coming from artificial intelligence without proper human oversight. Use of intelligent machines as instruments of aggression should also not be left to the machines themselves without proper human supervision. Eventually, artificial intelligence will increase to the extent that machines will need their own system of ethics and moral judgment. How far we can hold on to the principle that humans have to make the final choice on critical issues—human ethics above machine ethics—remains to be seen, especially after the point of singularity, when machine intelligence exceeds human intelligence.

### 11.5.2 Enhancing Our Own Life and Lifespan

Thanks to advances in medical science and public health management, we are, in general, living longer lives, with fewer illnesses and better capability to take care of our physical and mental conditions in old age. War, poverty, and self-inflicted damage, such as that which comes from eating too much, drinking, or smoking, will still badly affect many, but a better quality of life can be expected for most of us in the decades ahead. Genomic and other diagnostic methods will provide us with base-line information on the status of our health, even before we are born, so that we can prevent or be prepared to deal with potential defects stemming from our genetic makeups. The field of personalized medicine will evolve to deal with the fact that we are all different through inheritance of unique sets of genes, which will determine our health and disease status through our lifestyles and interactions with the environment. Some genetic diseases can be prevented or alleviated through editing of the genes (see Box 7.2), and others through selection of foods, drugs, and the environment around us so as to avoid the risk of potentially dangerous interactions with some of the genes identified in advance. Genome technologies will enable stratification of the population through different characteristics (extension of blood typing to typing of all aspects of our biological characters) and enable new public health approaches based on genomes. This will enable populations to be prepared for the effects of their genes on their lives and modify their behavior and exposure to various environments and foods. It will have beneficial impacts on health care and disease prevention. But it also brings the risks of misuse, such as invasion of privacy when genomic information is used without authorization of the owner. Without due ethical considerations, genomic information can be the basis for profiling and discrimination in jobs, schooling, and various areas of social interactions.

Longevity and prevention and treatment of many diseases will be achieved through repair and rejuvenation of organs and life systems with the use of stem cells and other medical technologies. Research is on the way to delay and even reverse aging changes in our cells and organs, so we will be able to live not only longer but also with a quality of life undiminished from that in our prime as we get older. Some parts may come from cells and tissues from other people, or even other species, and some may come from smart materials with embedded intelligent engineering systems. In addition to living with artificial intelligence agents like robots, we will also become partly bionic, with robotic and other smart, artificial systems inside us. Again, a number of ethical and moral issues have to be dealt with in the coming of this development. How much can we allow artificial systems and materials to be parts of ourselves? Should we allow them to replace parts of our own brain and other neural systems? Where do we draw the line between letting ourselves die and holding on with ever more artificial interference? Who makes these decisions in the case that our brains are already so weakened as to not be able to make a judgment by ourselves? These and other questions need to be seriously tackled as we embark more and more into the "brave new world."

### 11.5.3 Improving Our Society and Environment

As discussed extensively in Chapter 10, the sparks from the spirit of science provide many inputs into various aspects of sustainability, both of the society and of the environment. We expect that they will contribute not only to mere sustainability, in the sense of simple continuation, but also to the improvement over present and past conditions. Apart from helping toward material well-being of the society, we expect to see a happier world, with fewer wars and conflicts that still plague our global society, and with greater equity among people with different backgrounds and economic statuses. We expect individuals to lead happier lives, not only from their economic well-being, but also from a sense of satisfaction with their own lives. Science is not the only factor in gearing us toward achievement of a better, more livable society, but it is an important contributing factor.

In the long-term future ahead, as we have hopefully successfully managed our present problems with the environment, such as climate change, pollution, and water and energy balance, we can also look further for new homes to accommodate the foreseen increase in the global population. We can explore the oceans, both the surface and underwater spaces, for new settlements. Utmost care and precautions should be taken in such undertakings, especially since they provide new habitats that are artificial in the sense that human beings are not naturally suited to live in them. We should also look further to outer space, such as Mars and beyond, for potential new settlements. With the help of sparks from the spirit of science, technical barriers may be finally overcome and allow us to build these new homes elsewhere in the universe. In any case, we should not just choose a solution to find new settlements in the oceans or to leave our planet when attempts to achieve sustainability fail, since this surely means that we will create problems anew wherever we go. We must therefore achieve solutions to our sustainability problems where we are living here, on earth, first.

## 11.6 Message to the Universe

Will we achieve sustainability and live in harmony with both ourselves and our environment eventually? The next few decades will be crucial, first in meeting the targets of sustainability, as agreed by the global community in 2030, and then in going to the era beyond mere sustainability. Science and technology—the spirit and its sparks—are crucial factors in the efforts. However, as has been stressed throughout this book, they are tools that must be used together with other human efforts, including those in social, political, economic, and business spheres. Importantly, a sense of living in harmony with the society and the environment has to be cultivated in people globally. These are difficult tasks, but not impossible.

Apart from our civilization, we do not know whether there are other extraterrestrial civilizations out there. Given the vast expanse of the universe (or even many parallel universes), there is a good possibility that other civilizations than our own exist out there somewhere. If we have a chance, how do we communicate with them, and what messages do we want them to have about our civilization? Naturally, we would like them to know about who we are and our culture, arts, literature, architecture, and the way we live in cities and elsewhere. We would like to introduce them to our beautiful planet with mountains and oceans and all the living beings in them. Naturally, we would like them to have good intentions toward us and our planet. The extraterrestrial civilizations will likely have vastly different forms and means of communications from us, and mutual understanding would be difficult, even if contact can be made. If we make an assumption that such extraterrestrial civilizations are populated with intelligent members, the only possible means of first communication would surely be through science and technology, with the universal language based on the rules of nature.

We have already tried to introduce ourselves to other intelligent members of the universe. The space probes Pioneer 10 and 11 sent by NASA in 1972 and 1973 to explore Jupiter and Saturn and the outer regions of the solar system eventually passed into deep space. They contain plaques made of gold-anodized aluminum with a message from humanity (Fig. 11.2). The message has a picture of a man and a woman, the space capsule, the position of the solar system and our planet, an explanation of the properties of hydrogen, and the binary number system. The suggestion for sending the message was made by the space scientist Carl Sagan and associates as an attempt for extraterrestrial life to discover our existence in this corner of the universe. The message was based as much as possible on the rules of science, which are presumably applicable anywhere in the universe. The space capsules have left the solar system, and the last message was received from Pioneer 10 in 2003, at a distance of 12 billion km from the earth. Pioneer 10 is headed for the constellation Taurus, and Pioneer 11 is headed for the constellation Scutum. Voyager 1 and 2 space missions, launched in 1977, also to study the planets and the space beyond, contain golden records with sound and images of life and culture on earth, again intentionally for any extraterrestrial intelligence to discover us. These are "messages in bottles" in the vast ocean of the universe and are unlikely to be discovered unless the extraterrestrial intelligence has vastly powerful means to detect them.

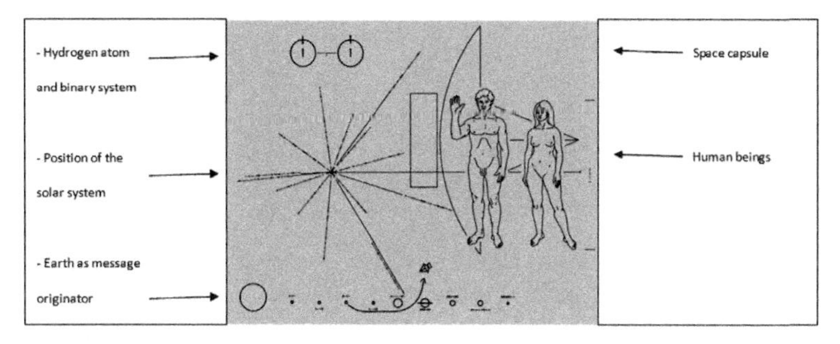


**Figure 11.2** Message from humanity to the universe, attached to space probes Pioneer 10 and 11. Vectors by Oona Raisanen (Mysid), design by Carl Sagan and Frank Drake, and artwork by Linda Saltzman Sagan. Public domain picture by NASA.

Rather than being discovered by extraterrestrial intelligence, the significance of the messages to the universe is more likely to act as time capsules, recording our own existence at this time in the history of the planet. We are like someone stranded on an isolated island in the middle of a vast ocean, crying out for someone to discover us. It is unlikely that we will be discovered. However, what if we are actually found? The irony will be that we are found only as a dead civilization on a dead planet.

Hopefully, we are found as a thriving civilization, ages and ages from now . . .

# Bibliography


1. Carson, R., *The Sense of Wonder*. 1998: Harper; Reprint Edition. 112 pp.
2. Bryson, B., *A Short History of Nearly Everything*. 2003: Broadway Books. 544 pp.
3. Hawking, S., and L. Mlodinow, *The Grand Design*. 1st ed. 2010: 198 pp.
4. Bohacek, R. S., C. McMartin, and W. C. Guida, The art and practice of structure-based drug design: a molecular modeling perspective. *Med. Res. Rev.*, **16**: pp. 3–50, 1996.
5. Yuthavong, Y., *Tapping the Molecular Wilderness*. 2016: Pan Stanford, Singapore. 134 pp.
6. Watson, J. D., and F. H. Crick, Molecular structure of nucleic acids; a structure for deoxyribose nucleic acid. *Nature*. **171**(4356): pp. 737–738, 1953.
7. Lovelock, J., *The Ages of Gaia. A Biography of Our Living Earth.* 2nd ed. 2000: Oxford University Press. 267 pp.
8. REN21, *Renewables 2015 Global Status Report*. 2015. http://www.ren21.net/status-of-renewables/global-status-report/.
9. IPCC, *Cimate Change 2013. The Physical Science Basis.* 2014: Cambridge University Press. 1552 pp.
10. Rockström, J., *Big World, Small Planet: Abundance within Planetary Boundaries*. 2015: Max Ström, Stockholm. 206 pp.
11. Eliot, T. S., *The Rock: A Pageant Play Written for Performance at Sadler's Wells Theatre 28 May - 9 June 1934 on behalf of the Forty-Five Church Funds of the Diocese*. 1934: Faber and Faber. 86 pp.
12. P21, *Partnership for 21st Century Learning*. 2016. http://www.p21.org/.
13. US National Research Council, *STEM Learning Is Everywhere: Summary of a Convocation on Building Learning Systems*, eds. S. Olson and J.

Labov (rapporteurs). 2014: National Academies Press, Washington DC. 90 pp.

14. Heckman, J., Presenting the Heckman equation, in *The Heckman Equation*. 2016. http://heckmanequation.org/blog.
15. Robinson, K., *Out of Our Minds. Learning to Be Creative*. 2011: Courier Westford, Westford, MA. 286 pp.
16. Einstein, A., *Einstein on Cosmic Religion and Other Opinions and Aphorisms*. 2009: Dover, New York. 97 pp.
17. Gibran, K., *The Prophet*. 1923: Alfred A. Knopf. 107 pp.
18. Gardner, H., *Frames of Mind. Theory of Multiple Intelligences*. 1983: Basic Books. 496 pp.
19. Doidge, N., *The Brain That Changes Itself*. 2007: Viking Press, New York. 427 pp.
20. Moravcsik, M. J., *How to Grow Science*. 1980: Universe Books, New York. 206 pp.
21. Chemical Heritage Foundation, *August Kekulé and Archibald Scott Couper*. 2016 (cited Mar. 1, 2016). http://www.chemheritage.org/discover/online-resources/chemistry-in-history/themes/molecular-synthesis-structure-and-bonding/kekule-and-couper.aspx.
22. Watson, J. D., *The Double Helix. A Personal Account of the Discovery of the Structure of DNA*. 1968. 256 pp.
23. Delistraty, C. C. *Can Creativity Be learned?* 2014. http://www.theatlantic.com/health/archive/2014/07/can-creativity-be-learned/372605/.
24. Latumahina, D. *9 Lessons Richard Feynman Taught Us about Creativity*. 2007. http://www.lifeoptimizer.org/2007/06/14/9-lessons-richard-feynman-taught-us-about-creativity/.
25. Jobs, S. *The Next Insanely Great Thing. The Wired Interview*. 1996. http://www.wired.com/1996/02/jobs-2/.
26. Kuhn, T., *The Structure of Scientific Revolutions*. 1962: University of Chicago Press. 264 pp.
27. Center for History of Physics, *Bright Idea: The First Laser. LaserFest, a Celebration of the 50th Anniversary of the Laser*. 2016. https://www.aip.org/history/exhibits/laser/.
28. American Chemical Society, *Discovery of Fullerenes National Historic Chemical Landmark*. 2010. http://www.acs.org/content/acs/en/education/whatischemistry/landmarks/fullerenes.html.
29. Godin, B., *Innovation Contested. The Ideas of Innovation over the Centuries*. 2015: Routledge, Oxford. 354 pp.

30. Ziman, J., A neural net model of innovation. *Sci. Public Policy*, **18**: pp. 65–75, 1991.
31. Goldacre, B., *Bad Science*. 2008: Fourth Estate, London. 370 pp.
32. Schumacher, E. F., *Small Is Beautiful. A Study of Economics As If People Mattered*. 1973: Blond&Briggs. 288 pp.
33. Horgan, J., *The End of Science: Facing the Limits of Knowledge in the Twilight of the Scientific Age*. 1996: Helix Books, Addison Wesley. 320 pp.
34. Brockman, J., *The Next Fifty Years*. 2002: Vintage Books.
35. Kurzweil, R., *The Singularity Is Near*. 2006: Viking.
36. Roco, M. C., et al., *Convergence of Knowledge, Technology and Society*. 2013: Springer, Heidelberg. 558 pp.
37. UNESCO Institute of Statistics, *Science, Technology and Innovation. Gross Domestic Expenditure on R&D (GERD)*. 2016. http://data.uis.unesco.org/.
38. World Health Organization, *World Malaria Report 2016*. 2016: Geneva. 186 pp.
39. World Health Organization, *Global Tuberculosis Report 2016*. 2016: Geneva. 201 pp.
40. Alexandratos, N., and J. Bruinsma, *World Agriculture towards 2030/2050: The 2012 Revision*. ESA working paper no. 12-03. 2012: Rome.
41. Prahalad, C. K., and S. L. Hart, *The Fortune at the Bottom of the Pyramid. Eradicating Poverty through Profits*. 2004: Wharton School, Philadelphia. 432 pp.
42. World Resources Institute and International Finance Corporation, *The Next Four Billion. Market Size and Business Strategy at the Base of the Pyramid*. 2007: World Resources Institute. 150 pp.
43. Avery, G. C., and H. Bergsteiner (eds.), *Sufficiency Thinking. Thailand's Gift to an Unsustainable World*. 2016: Allen and Unwin, Sydney. 293 pp.
44. Waldholz, M., and H. Pickersgill, *Grand Challenges in Global Health: 2005-2015*. 2015: Bill and Melinda Gates Foundation. 114 pp.
45. Cressey, D., Hippocratic oath for scientists, *Nature*. 2007.
46. Raup, D. M., and J. J. Sepkoski, Mass extinctions in the marine fossil record. *Science*. **215**: pp. 1501–1503, 1982.
47. Diamond, J., *Collapse*. 2005: Penguin Group, London. 575 pp.
48. Carson, R., *Silent spring*. 1962: Riverside Press, Boston. 368 pp.

49. Meadows, D. H., et al., *Limits to Growth*. 1972: Universe Books.
50. World Commission on Environment and Development, *Our Common Future*. 1987: Oxford University Press.
51. United Nations, *Earth Summit. UN Conference on Environment and Development (1992)*. 1997. http://www.un.org/geninfo/bp/enviro.html.
52. Grossman, N. (ed.), *Thailand's Sustainable Development Sourcebook*. 2015: Editions Didier Millet, Bangkok. 416 pp.
53. Baxter, W., N. Grossman, and N. Wegner (eds.), *A Call to Action. Thailand and the Sustainable Development Goals*. 2016: Editions Didier Millet, Bangkok. 183 pp.
54. Sachs, J. D., *The Age of Sustainable Development*. 2015: Columbia University Press, New York. 543 pp.
55. Motesharrei, S., J. Rivas, and E. Kalnay, Human and nature dynamics (HANDY): modeling inequality and use of resources in the collapse or sustainability of societies. *Ecol. Econ.*, **101**: pp. 90–102, 2014.
56. Sachs, J., et al., *SDG Index and Dashboards: Global Report*. 2016: Bertelsman Stiftung and Sustainable Development Solutions Network, New York.
57. Helliwell, J., R. Layard, and J. D. Sachs, *World Happiness Report 2016, Update (Vol. I)*. 2016: Sustainable Development Solutions Network, New York.
58. Jeffrey, K., H. Wheatley, and S. Abdallah, *The Happy Planet Index: 2016. A Global Index of Sustainable Well-Being*. 2016: New Economics Foundation, London.
59. Vinge, V., *The Coming Technological Singulaity: How to Survive in the Post-Human Era*, in *VISION 21 Symposium*. 1993.
60. Kurzweil, R., *The Singularity Is Near: When Humans Transcend Biology*. 2006: Viking Adult, 652 pp.

# Index